Springer Theses

Recognizing Outstanding Ph.D. Research

For further volumes:
http://www.springer.com/series/8790

Aims and Scope

The series "Springer Theses" brings together a selection of the very best Ph.D. theses from around the world and across the physical sciences. Nominated and endorsed by two recognized specialists, each published volume has been selected for its scientific excellence and the high impact of its contents for the pertinent field of research. For greater accessibility to non-specialists, the published versions include an extended introduction, as well as a foreword by the student's supervisor explaining the special relevance of the work for the field. As a whole, the series will provide a valuable resource both for newcomers to the research fields described, and for other scientists seeking detailed background information on special questions. Finally, it provides an accredited documentation of the valuable contributions made by today's younger generation of scientists.

Theses are accepted into the series by invited nomination only and must fulfill all of the following criteria.

- They must be written in good English.
- The topic of should fall within the confines of Chemistry, Physics and related interdisciplinary fields such as Materials, Nanoscience, Chemical Engineering, Complex Systems and Biophysics.
- The work reported in the thesis must represent a significant scientific advance.
- If the thesis includes previously published material, permission to reproduce this must be gained from the respective copyright holder.
- They must have been examined and passed during the 12 months prior to nomination.
- Each thesis should include a foreword by the supervisor outlining the significance of its content.
- The theses should have a clearly defined structure including an introduction accessible to scientists not expert in that particular field.

Albert Bartók-Pártay

The Gaussian Approximation Potential

An Interatomic Potential Derived from First Principles Quantum Mechanics

Doctoral Thesis accepted
by University of Cambridge, UK

Springer

Author
Albert Bartók-Pártay
Cavendish Laboratory, TCM Group
University of Cambridge
19, J J Thomson Avenue
CB3 0HE Cambridge, UK
ab686@cam.ac.uk

Supervisor
Prof. Mike Payne
Cavendish Laboratory, Head of TCM Group
University of Cambridge
19, J J Thomson Avenue
CB3 0HE Cambridge, UK
mcp1@cam.ac.uk

ISSN 2190-5053 e-ISSN 2190-5061

ISBN 978-3-642-14066-2 e-ISBN 978-3-642-14067-9

DOI 10.1007/978-3-642-14067-9

Springer Heidelberg Dordrecht London New York

Library of Congress Control Number: 2010930292

Cover Design: eStudio Calamar, Berlin/Figueres

Printed on acid-free paper

Springer is part of Springer Science+Business Media (www.springer.com)

Supervisor's Foreword

Atomistic simulations play an increasingly important role in many branches of science and, in the future, they will play a crucial role in understanding complex processes in biology and in developing new materials to confrontchallenges concerning energy and sustainability. The most accurate simulations use quantum mechanics to determine the interactions between the atoms. However, such approaches are restricted to moderate numbers of atoms, certainly less than a million, and, perhaps more importantly, very restricted timescales, generally less than a nanosecond and more likely of the order of picoseconds for the largest systems. Given these restrictions, then it is clear that the most complex problems in materials and biology will never be addressed using quantum mechanics alone. In contrast, atomistic simulations based on empirical interatomic potentials are computationally inexpensive. They have been used for simulations of systems containing billions of atoms for timescales of microseconds or even longer. Good empirical interatomic potentials exist for metals and for ionic materials but even these fail to describe all atomic environments with equal accuracy. Creating accurate interatomic potentials for covalent materials has proved to be extremely difficult.

The generation of interatomic potentials has, historically, been an art rather than an exact science. Furthermore, all interatomic potentials have limited regions of applicability though it has not been possible to quantify these regions in a useful way in order to prevent simulations performed with such potentials producing unphysical results. The new scheme for creating interatomic potentials described in this thesis generates accurate forces for any new atomic configuration based on a database of configurations and corresponding forces and energies gathered from any source although, realistically, the most useful source is likely to be quantum mechanical calculations. The new scheme does not fit an explicitly parameterised model of the interatomic potential but instead uses a Gaussian Process to determine the forces on the atoms. Indeed, one of the most interesting conclusions of this approach is that the use of parameterised models for interatomic potentials is, itself, responsible for most of the severe problems encountered in the development of interatomic potentials. This new approach to describing interatomic interactions is called the Gaussian Approximation Potential.

The key mathematical advance that made this approach possible was the description of the atomic neighbourhood in terms of the elements of the bispectrum. The bispectrum provides a very efficient description of the local atomic structure that takes account of the symmetries of the system, namely rotational, permutational and translational symmetries. Hence the bispectrum is invariant under rotations, translations and permutations of the sets atoms surrounding any particular atom. The bispectrum is a very effective descriptor of local structure and it will be interesting to see whether it will be applied to other problems, such as phase transitions, in the future. Importantly, the Gaussian Process also provides an estimate of the errors in the forces so simulations based on this potential can avoid the problem of generating spurious unphysical results due to errors in the interatomic potential. This situation would only be encountered if the simulation visited atomic configurations which were distant from the ones used to generate the original potential. One of the attractive features of the new approach is that, at this point, a new potential could be generated incorporating these new atomic configurations and the simulation could be continuedusing this new potential. This procedure would not reduce the accuracy of the potential for atomic configurations close to those in the original fitting set. This is not the case when reparameterising conventional interatomic potentials. Thus, Gaussian Approximation Potentials maintain accuracy as their transferability to new atomic environments is enhanced. In contrast, standard parameterised interatomic potentials always involve a compromise between accuracy and transferability.

These new potentials will also solve a number of the technical problems that presently exist in hybrid or QM/MM modelling schemes in which one or more parts of a system are described using quantum mechanics and the remainder of the system is described using simple analytical empirical potentials. The properties of the regions described using these new potentials would accurately match the properties of the regions described quantum mechanically thus reducing the problems introduced by the mismatch of, say, elastic properties across the interface. Furthermore, by continually adding quantum mechanical data to the training set for the potential, the Gaussian Approximation Potential could replace explicit quantum mechanical calculations for the quantum regions in such simulations yet maintain the accuracy of explicit quantum mechanical calculations. This approach would have a profound effect of significantly increasing the timescales that are accessible to QM/MM simulations. As a result of the work described in the thesis, there is now a standard procedure to determine accurate forces to use in any atomistic simulations. The Gaussian Approximation Potentials presented in the thesis for gallium nitride and for iron were developed in a few days (which included the generation of the input quantum mechanical data), yet are more accurate than existing potentials that were developed over a much longer period of time. In the future, I would expect that the whole process of generating GAP could be automated to allow any user to apply the method to any system of interest.

Cambridge, May 2010 Mike Payne

Acknowledgements

I am most grateful to my supervisor, Gábor Csányi, for his advice, help and discussions, which often happened outside office hours. I would also like to thank Mike Payne for giving the opportunity to carry out my research in the Theory of Condensed Matter Group. Thanks are due to Risi Kondor, whose advice on Group Theory was invaluable. I am especially thankful to Edward Snelson, who gave useful advice on the details of machine learning algorithms. Further thanks are due to my second supervisor, Mark Warner.

I am indebted to the members of Gábor Csány's research group: Lívia Bartók-Pártay, Noam Bernstein, James Kermode, Anthony Leung, Wojciech Szlachta, Csilla Várnai and Steve Winfield for useful discussions and inspiration at our group meetings. I am grateful to my peers in TCM, in particular, Hatem Helal and Mikhail Kibalchenko, from whom I received support and company.

Many thanks to Michael Rutter for his advice on computer-related matters. Thanks also to Tracey Ingham for helping with the administration.

I would like to thank my family for their patience and my wife, Lívia Bartók-Pártay for her love and encouragement.

Cambridge, May 2010 Albert Bartók-Pártay

Contents

Chapter 1
Introduction

Understanding the behaviour of materials at the atomic scale is fundamental to modern science and technology. As many properties and phenomena are ultimately controlled by the details of the atomic interactions, simulations of atomic systems provide useful information, which is often not accessible by experiment alone. Observing materials on a microscopic level can help to interpret physical phenomena and to predict the properties of previously unknown molecules and materials. To perform such atomistic simulations, we have to use models to describe the atomic interactions, whose accuracy has to be validated in order to ensure that the simulations are realistic.

Quantum Mechanics provides a description of matter, which, according to our current knowledge, is ultimately correct, a conclusion which is strongly corroborated by experimental evidence. However, the solution of the Schrödinger equation—apart from a few very simple examples—has to be performed numerically using computers. A series of approximations and sophisticated numerical techniques has led to various implementations of the originally exact quantum mechanical theory, which can be now routinely used in studies of atomic systems. In the last few decades, as computational speed capacities grew exponentially, the description of more and more atoms has become tractable. In most practical applications, the electrons and the nuclei are treated separately, and the quantum mechanical description of the nuclei is dropped altogether. This simplification, namely, that the nuclei move on a potential energy surface determined by the interaction of the electrons, already makes quantum mechanical calculations several order of magnitudes faster. However, determining macroscopic thermodynamical quantities of atomic systems requires a large number of samples of different arrangements of atoms, and the number of atoms has to be large enough to minimise finite-size effects. In fact, the computational costs associated with the solution of the Schrödinger equation are so large that the use of Quantum Mechanics is limited at most to a hundred of atoms and only a small fraction of the available configurational space.

The demand for faster calculations to allow calculations of larger systems or the exploration of configurational space leads to the realm of analytical potentials,

A. Bartók-Pártay, *The Gaussian Approximation Potential*, Springer Theses,
DOI: 10.1007/978-3-642-14067-9_1, © Springer-Verlag Berlin Heidelberg 2010

which are based on substituting the solution of the electronic Schrödinger equation with evaluation using an analytic function. Whereas the quantum mechanical description does not need validation—apart from ensuring that the errors introduced by the approximations are minimised—analytic potentials have to be checked to determine whether the description remains valid. This is often done by comparing macroscopic quantities computed by the model to experimental values. There is a high degree of arbitrariness in the creation and validation of such potentials [1], and in practice it is found that they are significantly less accurate than Quantum Mechanics.

As quantum mechanical calculations are becoming more widely available via computer programs such as CASTEP [2], CP2K [3] or Gaussian [4], we have access to a large number of microscopic observables. The approach we present in this thesis is to create interatomic potentials based directly on quantum mechanical data which are fast and have an accuracy close to the original method. To achieve this, we have used a Gaussian Process to interpolate the quantum mechanical potential energy surface. The Gaussian Process is routinely used by the machine-learning community for regression, but it has never previously been adapted to represent the atomic potential energy surface.

In traditional parametric regression a fixed functional form is assumed and a number of free parameters are optimised such that the resulting function is the best representation of the data. In case of complex, multivariate functions it is difficult to find suitable functional forms and it can be often observed that refitting with additional data significantly worsens the quality of the fit. On the other hand, non-parametric, nonlinear regression methods such as Gaussian Processes and neural networks offer flexible function fitting, with no prior assumptions for the functional form. Adding new, complementary data does not affect the quality of the fit, although this feature also suggests that the extrapolation capabilities of these methods are limited. Excellent introductions to machine learning methods can be found in [5] and [6].

Another important component of our method is the representation of atomic environments. We describe the environment of the atoms by a vector, called the bispectrum, which is invariant to rotations, translations and permutation of atoms in the neighbourhood. The bispectrum has been used in signal processing originally, but Kakarala generalised the concept [7] and Kondor derived the explicit formulae for the rotational groups. The well-known bond order parameters [8] are, in fact, a subset of the bispectrum. Within the bispectrum representation, we regard the potential energy surface as a sum of atomic energy functions, whose variables are the elements of the bispectrum.

This approach for generating interatomic potentials, which we collectively refer to as Gaussian Approximation Potentials, has the favourable scaling and speed of analytic potentials, while the accuracy is comparable with the underlying quantum mechanical method. With Gaussian Approximation Potentials atomistic simulations can be taken to an entirely new level.

1.1 Outline of the Thesis

The thesis is organised as follows. In Chap. 2 I discuss the representation of atomic environments by the bispectrum. I show how the rotational invariance of the bispectrum can be proved using Representation Theory and how the bispectrum is related to the widely used bond-order parameters. I summarise the Gaussian Process nonlinear regression method we used in Chap. 3, where I show the derivation of the formulae based on the Bayes' Theorem and the extensions which allowed us to use Gaussian Process for the regression of atomic potential energy surfaces. I describe a number of interatomic potentials and the Gaussian Approximation Potential in detail in Chap. 4. Details of the computational methods, which we used to test our model, are given in Chap. 5. Finally, I present our results on generating Gaussian Approximation Potentials for several systems and the validation of the models in Chap. 6.

References

1. D.W. Brenner, Phys. Stat. Sol. (b) **217**, 23 (2000)
2. S.J. Clark et al., Zeit. Krist. **220**, 567 (2005)
3. J. VandeVondele et al., Comp. Phys. Commun. **167**, 103 (2005)
4. M.J. Frisch et al., *Gaussian 09 Revision a.1.* (Gaussian Inc., Wallingford, CT, 2009)
5. D.J.C. MacKay, *Information Theory, Inference, and Learning Algorithms* (Cambridge University Press, Cambridge, UK, 2003), http://www.inference.phy.cam.ac.uk/mackay/itila/
6. C.E. Rasmussen, C.K.I. Williams, *Gaussian Processes for Machine Learning Adaptive Computation and Machine Learning* (MIT Press, Cambridge, MA, USA, 2006)
7. R. Kakarala, Triple Correlation on Groups, PhD thesis, Department of Mathematics, University of California, Irvine, 1992
8. P.J. Steinhardt, D.R. Nelson, M. Ronchetti, Phys. Rev. B **28**, 784 (1983)

Chapter 2
Representation of Atomic Environments

2.1 Introduction

The quantitative representation of atomic environments is an important tool in modern computational chemistry and condensed matter physics. For example, in structure search applications [1], each configuration that is found during the procedure depends numerically on the precise initial conditions and the path of the search, so it is important to be able to identify equivalent structures or detect similarities. In other applications, such as molecular dynamics simulation of phase transitions [2], one needs good order parameters capable of detecting changes in the local order around the atoms. In constructing interatomic potentials [3], the functional forms depend on elements of a carefully chosen representation of atomic neighbourhoods, e.g. bond lengths, bond angles, etc.

Although the Cartesian coordinate system provides a simple and unequivocal description of atomic systems, comparisons of structures based on it are difficult: the list of coordinates can be ordered arbitrarily, or two structures might be mapped to each other by a rotation, reflection or translation. Hence, two different lists of atomic coordinates can in fact represent the same or very similar structures. In a good representation, permutational, rotational and translational symmetries are built in explicitly, i.e. the representation is *invariant* with respect to these symmetries, while retaining the faithfulness of the Cartesian coordinates. If a representation is *complete*, a one-to-one mapping is obtained between the genuinely different atomic environments and the set of invariants comprising the representation.

The most well known invariants describing atomic neighbourhoods are the set of bond-order parameters proposed by Steinhardt et al. [4]. These have been successfully used as order parameters in studies of nucleation [5], phase transitions [6] and glasses [7]. In the following sections we show that the bond-order parameters actually form a subset of a more general set of invariants called the bispectrum. We prove that the bispectrum components indeed form a rotational and permutational invariant representation of atomic environments. The formally

A. Bartók-Pártay, *The Gaussian Approximation Potential*, Springer Theses,
DOI: 10.1007/978-3-642-14067-9_2, © Springer-Verlag Berlin Heidelberg 2010

infinite array of bispectral invariants provide an almost complete set, and by truncating it one obtains representations whose sensitivity can be refined at will.

2.2 Translational Invariants

The concept of the power spectrum and the bispectrum was originally introduced by the signal processing community. In the analysis of periodic signals the absolute phase is often irrelevant and a hindering factor, for example, when comparing signals. The problem of eliminating the phase of a periodic function is very similar to the problem of creating a rotationally invariant representation of spatial functions. We show how the bispectrum of periodic functions can be defined and discuss its possible uses in atomistic simulations.

2.2.1 Spectra of Signals

A periodic signal $f(t)$ (or a function defined on the circumference of a circle) where $t \in [0, 2\pi)$, can be represented by its Fourier series:

$$f(t) = \sum_n f_n \exp(i\omega_n t), \tag{2.1}$$

where the coefficients, f_n, can be obtained as follows:

$$f_n = \frac{1}{2\pi} \int_0^{2\pi} f(t) \exp(-i\omega_n t) \mathrm{d}t. \tag{2.2}$$

A phase shift of the signal (or rotation of the function) by t_0 transforms the original signal according to

$$f(t) \to f(t + t_0), \tag{2.3}$$

and the coefficients become

$$f_n \to \exp(i\omega_n t_0) f_n. \tag{2.4}$$

It follows that the power spectrum of the signal defined as

$$p_n = f_n^* f_n \tag{2.5}$$

is invariant to such phase shifts:

$$p_n = f_n^* f_n \to (f_n \exp(i\omega_n t_0))^* (f_n \exp(i\omega_n t_0)) = f_n^* f_n, \tag{2.6}$$

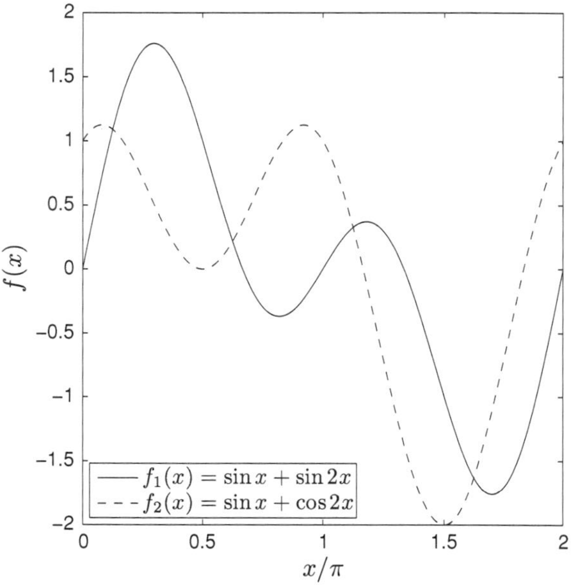

Fig. 2.1 Two different periodic functions that share the same power spectrum coefficients

Table 2.1 Fourier and power spectrum coefficients of f_1 and f_2

ω	−2	−1	0	1	2
$\mathbf{f}_1$	$-i$	$-i$	0	i	i
$\mathbf{f}_2$	1	$-i$	0	i	1
$\mathbf{p}_1 = \mathbf{p}_2$	1	1	0	1	1

but the information content of different channels becomes decoupled. Figure 2.1 and Table 2.1 demonstrate two functions, $f_1 = \sin(t) + \sin(2t)$ and $f_2 = \sin(t) + \cos(2t)$, that can both be represented by the same power spectrum.

2.2.2 *Bispectrum*

As the power spectrum is not complete, i.e. the original function cannot be reconstructed from it, there is a need for an invariant representation from which the original function can (at least in theory) be restored. The bispectrum contains the relative phase of the different channels, moreover, it has been proven to be complete [8].

A periodic function $f : \mathbb{R}^n \to \mathbb{C}$, whose period is L_i in the ith direction, can be expressed in terms of a Fourier series:

$$f(\mathbf{r}) = \sum_{\omega} f(\omega) \exp(i\omega\mathbf{r}), \tag{2.7}$$

where the Fourier-components can be obtained from

$$f(\boldsymbol{\omega}) = \prod_{i=1}^{n} \frac{1}{L_i} \int_V f(\mathbf{r}) \exp(i\boldsymbol{\omega}\mathbf{r}) d\mathbf{r} \tag{2.8}$$

and $\boldsymbol{\omega} = (\omega_1, \omega_2, \ldots, \omega_n)$. An arbitrary translation $\hat{T}(\mathbf{r}_0)$ transforms f as $f(\mathbf{r}) \to f(\mathbf{r} - \mathbf{r}_0)$, thus the Fourier-coefficients change as $f(\boldsymbol{\omega}) \to \exp(-i\boldsymbol{\omega}\mathbf{r}_0)f(\boldsymbol{\omega})$. The bispectrum of f is defined as the triple-correlation of the Fourier coefficients:

$$b(\boldsymbol{\omega}_1, \boldsymbol{\omega}_2) = f(\boldsymbol{\omega}_1)f(\boldsymbol{\omega}_2)f(\boldsymbol{\omega}_1 + \boldsymbol{\omega}_2)^*. \tag{2.9}$$

The bispectrum is invariant to translations:

$$\begin{aligned} b(\boldsymbol{\omega}_1, \boldsymbol{\omega}_2) \to & f(\boldsymbol{\omega}_1) \exp(i - \boldsymbol{\omega}_1 \mathbf{r}_0) f(\boldsymbol{\omega}_2) \exp(i - \boldsymbol{\omega}_2 \mathbf{r}_0) \\ & \times f(\boldsymbol{\omega}_1 + \boldsymbol{\omega}_2)^* \exp(i(\boldsymbol{\omega}_1 + \boldsymbol{\omega}_2)\mathbf{r}_0) = b(\boldsymbol{\omega}_1, \boldsymbol{\omega}_2). \end{aligned} \tag{2.10}$$

The bispectrum has been shown to be complete [8]. The proof, which is highly technical and would be too long to reproduce here is based on Group Theory. Further, Dianat and Raghuveer [9] proved that in case of one- and two-dimensional functions the original function can be restored using only the diagonal elements of the bispectrum, i.e. only the components for which $\boldsymbol{\omega}_1 = \boldsymbol{\omega}_2$.

2.2.3 *Bispectrum of Crystals*

Crystals are periodic repetitions of a unit cell in space in each of the three directions defined by the lattice vectors. A unit cell can be described as a parallelepiped (the description used by the conventional Bravais system of lattices) containing some number of atoms at given positions. The three independent edges of the parallelepiped are the lattice vectors, whereas the positions of the atoms in the unit cell form the basis. Defining crystals in this way is not unique, as any subset of a crystal which generates it by translations can be defined as a unit cell, for example, a Wigner–Seitz cell, which is not even necessarily a parallelepiped.

Thus a crystal can be described by the coordinates of the basis atoms $\mathbf{r}_i$, where $i = 1, \ldots, N$ and the three lattice vectors $\mathbf{a}_{\alpha}, \alpha = 1, 2, 3$. The position of the basis can be given in terms of the fractional coordinates $\mathbf{x}_i$, such that

$$\mathbf{r}_i = \sum_{\alpha=1}^{3} x_{i\alpha} \mathbf{a}_{\alpha}, \tag{2.11}$$

where $0 < x_{i\alpha} < 1$.

In the same way as in the case of atomic environments, the order of the atoms in the basis is arbitrary. We introduce the permutational invariance through the atomic density:

$$\rho(\mathbf{x}) = \sum_i \delta(\mathbf{x} - \mathbf{x}_i). \tag{2.12}$$

ρ is a periodic function in the unit cube, therefore we can expand it in a Fourier series and calculate invariant features such as the power spectrum and bispectrum. It can be noted that the power spectrum of ρ is equivalent to the structure factor used in X-ray and neutron diffraction, and it is clear from Sect. 2.2.1 why the structure factor is not sufficient to determine the exact structure of a crystal. In contrast, the bispectrum of the atomic density function could be used as a unique fingerprint of the crystal that is invariant to the permutation and translation of the basis.

We note that permuting the lattice vectors of the crystal permutes the reciprocal lattice vectors which therefore, mixes the elements of the bispectrum. This problem can be eliminated by first matching the lattice vectors of the two structures which are being compared. The rotation of the entire lattice does not change the fractional coordinates, hence the bispectrum is invariant to global rotations.

2.3 Rotationally Invariant Features

Invariant features of atomic environments can be constructed by several methods, of which we list a few here. In interatomic potentials, a set of geometric parameters are used, such as bond lengths, bond angles and tetrahedral angles. These are rotationally invariant by construction, but the size of a complete set of such parameters grows as exp(N), where N is the number of neighbours. The complete set is vastly redundant, but there is no systematic way of reducing the number of parameters without losing completeness.

A more compact rotationally invariant representation of the atomic environment can be built in the form of a matrix by using the bond vectors $\mathbf{r}_i, i = 1,\ldots,N$ between the central atom and its N neighbours. The elements of the matrix are given by the dot product

$$M_{ij} = \mathbf{r}_i \cdot \mathbf{r}_j. \tag{2.13}$$

Matrix $\mathbf{M}$ contains the bond lengths on its diagonal, whereas the off-diagonal elements are related to the bond angles. It can be shown that $\mathbf{M}$ is a complete representation [10]. However, permuting the neighbouring atoms shuffles the columns and rows of $\mathbf{M}$, thus $\mathbf{M}$ is not a suitable invariant representation.

Permutational invariance can be achieved by using the symmetric polynomials [11]. These are defined by

$$\Pi_k(x_1, x_2, \ldots, x_N) = \Pi_k(x_{\pi_1}, x_{\pi_2}, \ldots, x_{\pi_N}) \tag{2.14}$$

for every $\boldsymbol{\pi}$, where $\boldsymbol{\pi}$ is an arbitrary permutation of the vector $(1, 2, \ldots, N)$. The first three symmetric polynomials are

$$\Pi_1(x_1, x_2, \ldots, x_N) = \sum_i^N x_i \tag{2.15}$$

$$\Pi_2(x_1, x_2, \ldots, x_N) = \sum_{i<j}^{N} x_i x_j \tag{2.16}$$

$$\Pi_3(x_1, x_2, \ldots, x_N) = \sum_{i<j<k}^{N} x_i x_j x_k. \tag{2.17}$$

The series of polynomials form a complete representation, however, this set is not rotationally invariant.

2.3.1 Bond-order Parameters

As a first step to derive a more general invariant representation of atomic environments, we define the local atomic density as

$$\rho_i(\mathbf{r}) = \sum_j \delta(\mathbf{r} - \mathbf{r}_{ij}), \tag{2.18}$$

where the index j runs over the neighbours of atom i. The local atomic density is already invariant to permuting neighbours, as changing the order of the atoms in the neighbour list only affects the order of the summation. This function could be expanded in terms of spherical harmonics (dropping the atomic index i for clarity):

$$\rho(\mathbf{r}) = \sum_{l=0} \sum_{m=-l}^{l} c_{lm} Y_{lm}(\theta(\mathbf{r}), \phi(\mathbf{r})). \tag{2.19}$$

However, we should note that this representation does not contain information about the distances of neighbours. In fact, $\rho(\mathbf{r})$ represented this way is the projection of the positions of neighbouring atoms onto the unit sphere. The properties of functions defined on the unit sphere are described by the group theory of SO(3), the group of rotations about the origin.

The spherical harmonics functions form an orthonormal basis set for L_2:

$$\langle Y_{lm} | Y_{l'm'} \rangle = \delta_{ll'} \delta_{mm'}, \tag{2.20}$$

where the inner product of functions f and g is defined as

$$\langle f | g \rangle = \int f^*(\mathbf{r}) g(\mathbf{r}) d\mathbf{r}. \tag{2.21}$$

The coefficients c_{lm} can be determined as

$$c_{lm} = \langle \rho | Y_{lm} \rangle = \sum_j Y_{lm}\big(\theta(\mathbf{r}_{ij}), \phi(\mathbf{r}_{ij})\big). \tag{2.22}$$

We note that the order parameters Q_{lm} introduced by Steinhardt et al. [4] are proportional to the coefficients c_{lm}. In their work, they defined the bonds in the system as vectors joining neighbouring atoms. Defining which atoms are the neighbours of a particular atom can be done by using a simple distance cutoff or via the Voronoi analysis. Once the set of neighbours has been defined, each bond $\mathbf{r}_{ij}$ connecting neighbour atoms i and j is represented by a set of spherical harmonics coefficients

$$Y_{lm}(\hat{\mathbf{r}}_{ij}) = Y_{lm}(\theta(\mathbf{r}_{ij}), \phi(\mathbf{r}_{ij})). \tag{2.23}$$

Averaging the coefficients for atom i provides the atomic order parameters for that atom

$$Q^i_{lm} = \frac{1}{N_i}\sum_j Y_{lm}(\hat{\mathbf{r}}_{ij}), \tag{2.24}$$

where N_i is the number of neighbours of atom i. Similarly, averaging over all bonds in the system gives a set of global order parameters

$$\bar{Q}_{lm} = \frac{1}{N_b}\sum_{ij} Y_{lm}(\hat{\mathbf{r}}_{ij}), \tag{2.25}$$

where N_b is the total number of bonds. Both of these order parameters are invariant to permutations of atoms and to translations, but they still depend on the orientation of the reference frame. However, rotationally invariant combinations of these order parameters can be constructed as follows:

$$Q^i_l = \left(\frac{4\pi}{2l+1}\sum_{m=-l}^{l}(Q^i_{lm})^* Q^i_{lm}\right)^{1/2} \quad \text{and} \tag{2.26}$$

$$W^i_l = \sum_{m_1,m_2,m_3=-l}^{l}\begin{pmatrix} l & l & l \\ m_1 & m_2 & m_3 \end{pmatrix} Q^i_{lm_1} Q^i_{lm_2} Q^i_{lm_2} \tag{2.27}$$

for atoms and

$$\bar{Q}_l = \left(\frac{4\pi}{2l+1}\sum_{m=-l}^{l}\bar{Q}^*_{lm}\bar{Q}_{lm}\right)^{1/2} \tag{2.28}$$

$$\bar{W}_l = \sum_{m_1,m_2,m_3=-l}^{l}\begin{pmatrix} l & l & l \\ m_1 & m_2 & m_3 \end{pmatrix} \bar{Q}_{lm_1}\bar{Q}_{lm_2}\bar{Q}_{lm_2} \tag{2.29}$$

for global structures. The factor in parentheses is the Wigner-3jm symbol, which is nonzero only for $m_1 + m_2 + m_3 = 0$.

Q^i_l and W^i_l are called second-order and third-order bond-order parameters, respectively. It is possible to normalise W^i_l such that it does not depend strongly on the number of neighbours as follows:

$$\hat{W}_l^i = W_l^i \Big/ \left(\sum_{m=-l}^{l} (Q_{lm}^i)^* Q_{lm}^i \right)^{3/2}. \tag{2.30}$$

Bond-order parameters were originally introduced by Steinhardt et al. [4] for studying the order in liquids and glasses, but their approach was adopted soon for a wide range of applications. For example, the bond-order parameters, when averaged over all bonds in the system, can be used as reaction coordinates in phase transitions [12].

For symmetry reasons, bond order parameters with $l \geq 4$ have non-zero values in clusters with cubic symmetry and $l \geq 6$ for clusters with icosahedral symmetry. The most widely calculated bond order parameters are $l = 4$ and $l = 6$. Different values correspond to crystalline materials with different symmetry, while the global values vanish in disordered phases, such as in liquids. This feature made the Q and W invariants attractive for use as bond order parameters in many applications.

2.3.2 Power Spectrum

Using some basic concepts from representation theory, we can now prove that the second-order invariants are rotationally invariant, then we show a more general form of invariants, a superset consisting of third-order invariants [13]. An arbitrary rotation $\hat{R}$ operating on a spherical harmonic function Y_{lm} transforms it into a linear combination of spherical harmonics with the same l index:

$$\hat{R} Y_{lm} = \sum_{m'=-l}^{l} D_{mm'}^{(l)}(R) Y_{lm'}, \tag{2.31}$$

where the matrices $\mathbf{D}^{(l)}(R)$ are also known as the Wigner-matrices. The elements of the Wigner matrices can be generated by

$$D_{mm'}^{(l)}(R) = \langle Y_{lm} | \hat{R} | Y_{lm'} \rangle. \tag{2.32}$$

It follows that the rotation operator $\hat{R}$ acts on the function ρ as

$$\begin{aligned} \hat{R}\rho &= \hat{R} \sum_{l=0} \sum_{m=-l}^{l} c_{lm} Y_{lm} = \sum_{l=0} \sum_{m=-l}^{l} c_{lm} \hat{R} Y_{lm} \\ &= \sum_{l=0} \sum_{m=-l}^{l} \sum_{m'=-l}^{l} c_{lm} D_{mm'}^{(l)}(R) Y_{lm'} \\ &= \sum_{l=0} \sum_{m'=-l}^{l} c'_{lm} Y_{lm'}, \end{aligned} \tag{2.33}$$

thus the vector of coefficients $\mathbf{c}_l$ transform under rotation as

$$\mathbf{c}_l \rightarrow \mathbf{D}^{(l)}(R)\mathbf{c}_l. \tag{2.34}$$

Making use of the fact that rotations are unitary operations, it is possible to show that the matrices $\mathbf{D}^{(l)}$ are unitary, i.e.

$$\left(\mathbf{D}^{(l)}\right)^{\dagger}\mathbf{D}^{(l)} = \mathbf{I}, \tag{2.35}$$

leading us to a set of rotationally invariant coefficients, the rotational power spectrum:

$$p_l = \mathbf{c}_l^{\dagger}\mathbf{c}_l. \tag{2.36}$$

The coefficients of the power spectrum remain invariant under rotations:

$$p_l = \mathbf{c}_l^{\dagger}\mathbf{c}_l \rightarrow \left(\mathbf{c}_l^{\dagger}\left(\mathbf{D}^{(l)}\right)^{\dagger}\right)\left(\mathbf{D}^{(l)}\mathbf{c}_l\right) = \mathbf{c}_l^{\dagger}\mathbf{c}_l. \tag{2.37}$$

It can be directly seen that the second-order bond-order parameters are related to the power spectrum via the simple equation

$$Q_l = \left(\frac{4\pi}{2l+1}p_l\right)^{1/2}. \tag{2.38}$$

The power spectrum is a very impoverished representation of the original function ρ, because all p_l coefficients are rotationally invariant *independently*, i.e. different l channels are decoupled. This representation, although rotationally invariant, is, in turn, severely incomplete.

The incompleteness of the power spectrum can be demonstrated by the following example. Assuming a function f in the form

$$f(\hat{\mathbf{r}}) = \sum_{m=-l_1}^{l_1} \alpha_m Y_{l_1 m}(\hat{\mathbf{r}}) + \sum_{m=-l_2}^{l_2} \beta_m Y_{l_2 m}(\hat{\mathbf{r}}), \tag{2.39}$$

its power spectrum elements are $p_{l_1} = |\boldsymbol{\alpha}|^2$ and $p_{l_1} = |\boldsymbol{\beta}|^2$. Thus only the *length* of the vectors $\boldsymbol{\alpha}$ and $\boldsymbol{\beta}$ are constrained by the power spectrum, their relative orientation is lost, i.e. the information content of channels l_1 and l_2 becomes decoupled. Figure 2.2 shows two different angular functions, $f_1 = Y_{22} + Y_{2\text{-}2} + Y_{33} + Y_{3\text{-}3}$ and $f_2 = Y_{21} + Y_{2\text{-}1} + Y_{32} + Y_{3\text{-}2}$ that have the same power spectrum $p_2 = 2$ and $p_3 = 2$.

2.3.3 *Bispectrum*

We will now generalise the concept of the power spectrum in order to obtain a more complete set of invariants via the coupling of the different angular

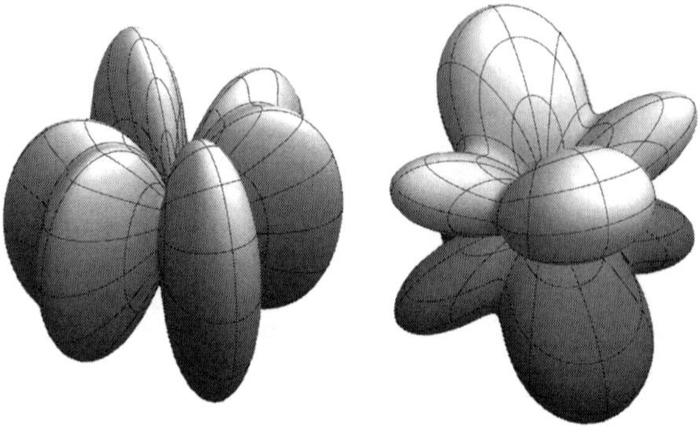

Fig. 2.2 Two different angular functions that share the same power spectrum coefficients

momentum channels [13]. Let us consider the direct product $\mathbf{c}_{l_1} \otimes \mathbf{c}_{l_2}$, which transforms under a rotation as

$$\mathbf{c}_{l_1} \otimes \mathbf{c}_{l_2} \rightarrow \left(\mathbf{D}^{(l_1)} \otimes \mathbf{D}^{(l_2)}\right)(\mathbf{c}_{l_1} \otimes \mathbf{c}_{l_2}). \tag{2.40}$$

It follows from the representation theory of groups that the direct product of two irreducible representations can be decomposed into direct sum of irreducible representations of the same group. In case of the SO(3) group, the direct product of two Wigner-matrices can be decomposed into a direct sum of Wigner-matrices in the form

$$\mathbf{D}^{(l_1)} \otimes \mathbf{D}^{(l_2)} = \left(\mathbf{C}^{l_1,l_2}\right)^\dagger \left[\bigoplus_{l=|l_1-l_2|}^{l_1+l_2} \mathbf{D}^{(l)}\right] \mathbf{C}^{l_1,l_2}, \tag{2.41}$$

where $\mathbf{C}^{l_1,l_2}$ denote the Clebsch–Gordan coefficients. The matrices of Clebsch–Gordan coefficients are themselves unitary, hence the vector $\mathbf{C}^{l_1,l_2}(\mathbf{c}_{l_1} \otimes \mathbf{c}_{l_2})$ transforms as

$$\mathbf{C}^{l_1,l_2}(\mathbf{c}_{l_1} \otimes \mathbf{c}_{l_2}) \rightarrow \left[\bigoplus_{l=|l_1-l_2|}^{l_1+l_2} \mathbf{D}^{(l)}\right] \mathbf{C}^{l_1,l_2}(\mathbf{c}_{l_1} \otimes \mathbf{c}_{l_2}). \tag{2.42}$$

We define $\mathbf{g}_{l_1,l_2,l}$ as

$$\bigoplus_{l=|l_1-l_2|}^{l_1+l_2} \mathbf{g}_{l_1,l_2,l} \equiv \mathbf{C}^{l_1,l_2}(\mathbf{c}_{l_1} \otimes \mathbf{c}_{l_2}), \tag{2.43}$$

i.e. the $\mathbf{g}_{l_1,l_2,l}$ is that part of the RHS which transforms under rotation as

$$\mathbf{g}_{l_1,l_2,l} \rightarrow \mathbf{D}^{(l)} \mathbf{g}_{l_1,l_2,l}. \tag{2.44}$$

Analogously to the power spectrum, the bispectrum components or cubic invariants, can be written as

$$b_{l_1,l_2,l} = \mathbf{c}_l^{\dagger} \mathbf{g}_{l_1,l_2,l}, \tag{2.45}$$

which are invariant to rotations:

$$b_{l_1,l_2,l} = \mathbf{c}_l^{\dagger} \mathbf{g}_{l_1,l_2,l} \rightarrow \left(\mathbf{c}_l \mathbf{D}^{(l)}\right)^{\dagger} \mathbf{D}^{(l)} \mathbf{g}_{l_1,l_2,l} = \mathbf{c}_l^{\dagger} \mathbf{g}_{l_1,l_2,l} \tag{2.46}$$

Kondor showed that the bispectrum of the SO(3) space is not complete, i.e. the bispectrum does not determine uniquely the original function. This is a deficiency due to the fact that the unit sphere, S_2 is a homogeneous space. However, he states that the bispectrum is still a remarkably rich invariant representation of the function.

Rewriting the bispectrum formula as

$$b_{l_1,l_2,l} = \sum_{m=-l}^{l} \sum_{m_1=-l_1}^{l_1} \sum_{m_2=-l_2}^{l_2} c_{lm}^{*} C_{l_1 m_1 l_2 m_2}^{lm} c_{l_1 m_1} c_{l_2 m_2}, \tag{2.47}$$

the similarity to the third-order bond-order parameters becomes apparent. Indeed, the Wigner 3jm-symbols are related to the Clebsch–Gordan coefficients through

$$\begin{pmatrix} l_1 & l_2 & l_3 \\ m_1 & m_2 & m_3 \end{pmatrix} = \frac{(-1)^{l_1-l_2-m_3}}{\sqrt{2l_3+1}} C_{l_1 m_1 l_2 m_2}^{lm}. \tag{2.48}$$

For the spherical harmonics $Y_{lm} = (-1)^m Y_{l-m}^{*}$, thus the third-order parameters W_l are simply the diagonal elements of the bispectrum $b_{l,l,l}$ up to a scalar factor, and thus, the bispectrum is a superset of the third-order bond-order parameters. Further, considering that $Y_{00} \equiv 1$, therefore the coefficient c_{00} is simply the number of neighbours N, and $C_{m,0,m_2}^{l,0,l_2} = \delta_{l,l_2} \delta_{m,m_2}$, we notice that the bispectrum elements $l_1 = 0,\ l\ = l_2$ are the power spectrum components, previously introduced:

$$b_{l,0,l} = N_i \sum_{m=-l}^{l} \sum_{m_2=-l}^{l} c_{lm}^{*} \delta_{m,m_2} c_{lm_2} = N_i \sum_{m=-l}^{l} c_{lm}^{*} c_{lm} = N_i p_l. \tag{2.49}$$

Finally, the relationship between the bond-order parameters and the bispectrum can be summarised as

$$Q_l \propto \sqrt{p_l} \propto \sqrt{b_{l,0,l}} \tag{2.50}$$

$$W_l \propto b_{l,l,l}. \tag{2.51}$$

2.3.3.1 Radial Dependence

The bispectrum is still a very incomplete representation, as it uses the unit-sphere projection of the atomic environment, i.e. the distance of the atoms from the centre is not represented. One way to improve this shortcoming—namely, the lack of radial information—is to introduce radial basis functions [14], completing the basis for three-dimensional space. In Eq. 2.19, we use the product of spherical harmonics and a linearly independent set of radial functions g_n:

$$\rho(\mathbf{r}) = \sum_n \sum_{l=0} \sum_{m=-l}^{l} c_{nlm} g_n(r) Y_{lm}(\theta(\mathbf{r}), \phi(\mathbf{r})). \tag{2.52}$$

If the set of radial basis functions is not orthonormal, i.e. $\langle g_n \mid g_m \rangle = S_{nm} \neq \delta_{nm}$, after obtaining the coefficients c'_{nlm} with

$$c'_{nlm} = \langle g_n Y_{lm} | \rho \rangle, \tag{2.53}$$

the elements c_{nlm} are given as

$$c_{nlm} = \sum_{n'} \left(S^{-1}\right)_{n'n} c'_{n'lm}. \tag{2.54}$$

In practice, when constructing the invariants, both c'_{nlm} and c_{nlm} can be used.

Rotational invariance only applies globally, therefore the different angular momentum channels corresponding to various radial basis functions need to be coupled. Simply extending Eq. 2.47 to the form

$$b_{n,l_1,l_2,l} = \sum_{m=-l}^{l} \sum_{m_1=-l_1}^{l_1} \sum_{m_2=-l_2}^{l_2} c^*_{nlm} C^{lm}_{l_1 m_1 l_2 m_2} c_{nl_1 m_1} c_{nl_2 m_2}, \tag{2.55}$$

provides a set of invariants describing the three-dimensional neighbourhood of the atom. In fact, this formula can easily lead to a poor representation, if the radial functions have little overlap with each other, as the coefficients belonging to different n channels become decoupled. To avoid this, it is necessary to choose wide, overlapping radial functions, although this greatly reduces the sensitivity of each channel (Fig. 2.3). The fine-tuning of the basis set is rather arbitrary, and there does not necessarily exist an optimum for all systems. An alternative way to construct invariants from $\mathbf{c}$ is to couple different radial channels, for example, as

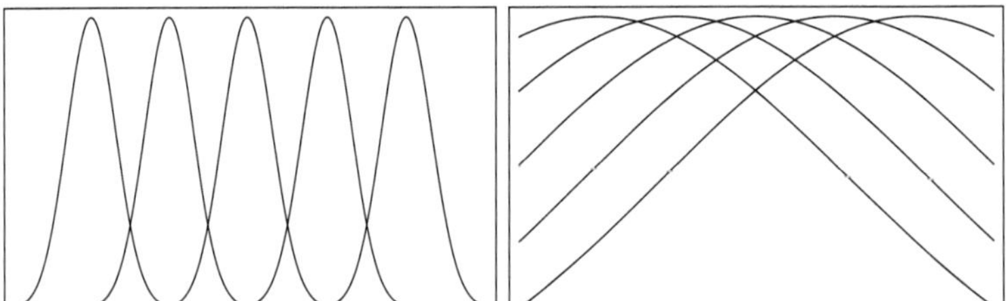

Fig. 2.3 Two possible sets of radial basis functions, Gaussians centred at different radii. The narrow Gaussians are more sensitive to changes in radial positions, but the coupling between them is weaker

$$b_{n_1,n_2,l_1,l_2,l} = \sum_{m=-l}^{l} \sum_{m_1=-l_1}^{l_1} \sum_{m_2=-l_2}^{l_2} c^*_{n_1 lm} C^{lm}_{l_1 m_1 l_2 m_2} c_{n_2 l_1 m_1} c_{n_2 l_2 m_2}. \tag{2.56}$$

Now we ensure that radial channels cannot become decoupled, but at the price of increasing the number of invariants quadratically. Although adding a suitable set of radial functions allows one to construct a complete representation, we found this approach overly complicated. A high degree of arbitrariness is introduced by having to choose a radial basis.

2.3.4 4-Dimensional Bispectrum

Instead of using a rather arbitrary radial basis set, we propose a generalisation of the power spectrum and bispectrum that does not require the explicit introduction of a radial basis set, yet still forms a complete basis of three-dimensional space. We start by projecting the atomic neighbourhood density onto the surface of the four-dimensional unit sphere, in a similar fashion to the Riemann-construction:

$$\mathbf{r} \equiv \begin{pmatrix} x \\ y \\ z \end{pmatrix} \to \begin{array}{l} \phi = \arctan(y/x) \\ \theta = \arccos(z/|\mathbf{r}|), \\ \theta_0 = |\mathbf{r}|/r_0 \end{array} \tag{2.57}$$

where $r_0 > r_{\text{cut}}/\pi$. Using this projection, rotations in the three-dimensional space correspond to rotations in the four-dimensional space. Figure 2.4 shows such projections for one and two dimensions, which can be more easily drawn than the three-dimensional case that we use here.

An arbitrary function ρ defined on the surface of a 4D sphere can be numerically represented using the hyperspherical harmonics functions $U^j_{m'm}(\phi, \theta, \theta_0)$:

$$\rho = \sum_{j=0}^{\infty} \sum_{m,m'=-j}^{j} c^j_{m'm} U^j_{m'm}. \tag{2.58}$$

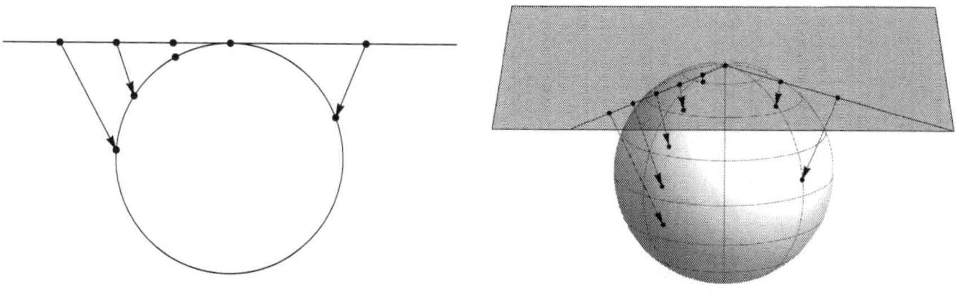

Fig. 2.4 Projection of a line to a circle (*left*), projection the two-dimensional plane onto the three-dimensional sphere (*right*). The projection we use is in Eq. 2.57 the generalisation to one more dimension

The hyperspherical harmonics form an orthonormal basis set, thus the expansion coefficients $c^j_{m'm}$ can be calculated via

$$c^j_{m'm} = \langle U^j_{m'm} | \rho \rangle, \tag{2.59}$$

where $\langle \cdot | \cdot \rangle$ denotes the inner product in four-dimensional space. Although the coefficients $c^j_{m'm}$ have two indices for each j, they are vectors and, for clarity, we denote them as $\mathbf{c}^j$. Similarly to the three-dimensional case, a unitary operation $\hat{R}$, such as a rotation, acts on the hyperspherical harmonics functions as

$$\hat{R} U^j_{m'_1 m_1} = \sum_{m'_2 m_2} R^j_{m'_1 m_1 m'_2 m_2} U^j_{m'_2 m_2}, \tag{2.60}$$

where the matrix elements $R^j_{m'_1 m_1 m'_2 m_2}$ are given by

$$R^j_{m'_1 m_1 m'_2 m_2} = \left\langle U^j_{m'_1 m_1} | \hat{R} | U^j_{m'_2 m_2} \right\rangle. \tag{2.61}$$

Hence the rotation $\hat{R}$ acting on ρ transforms the coefficient vectors $\mathbf{c}^j$ according to

$$\mathbf{c}^j \rightarrow \mathbf{R}^j \mathbf{c}^j. \tag{2.62}$$

$\mathbf{R}^j$ are unitary matrices, i.e. $\left(\mathbf{R}^j\right)^\dagger \mathbf{R}^j = \mathbf{I}$.

The product of two hyperspherical harmonics functions can be expressed as the linear combination of hyperspherical harmonics [15]:

$$U^{l_1}_{m'_1 m_1} U^{l_2}_{m'_2 m_2} = \sum_{l=|l_1-l_2|}^{l_1+l_2} C^{lm}_{l_1 m_1 l_2 m_2} C^{lm}_{l_1 m_1 l_2 m_2} U^l_{m'm}, \tag{2.63}$$

where $C^{lm}_{l_1 m_1 l_2 m_2}$ are the well-known Clebsch–Gordan coefficients. We can recognise in Eq. 2.63 the four dimensional analogues of the Clebsch–Gordan expansion coefficients, defined as $H^{lmm'}_{l_1 m_1 m'_1, l_2 m_2 m'_2} \equiv C^{lm}_{l_1 m_1 l_2 m_2} C^{lm'}_{l_1 m'_1 l_2 m'_2}$. Using the matrix notation of the expansion coefficients, it can be shown that the direct product of the four-dimensional rotation matrices decompose according to

$$\mathbf{R}^{j_1} \otimes \mathbf{R}^{j_2} = \left(\mathbf{H}^{j_1 j_2}\right)^\dagger \left[\bigoplus_{j=|j_1-j_2|}^{j_1+j_2} \mathbf{R}^j \right] \mathbf{H}^{j_1 j_2}. \tag{2.64}$$

The remainder of the derivation continues analogously to the 3D case. Finally, we arrive at the expression for the bispectrum elements, given by

$$B_{j_1 j_2 j} = \sum_{m'_1, m_1 = -j_1}^{j_1} \sum_{m'_2, m_2 = -j_2}^{j_2} \sum_{m', m = -j}^{j} \left(c^j_{m'm}\right)^* C^{jm}_{j_1 m_1 j_2 m_2} C^{jm'}_{j_1 m'_1 j_2 m'_2} c^{j_1}_{m'_1 m_1} c^{j_2}_{m'_2 m_2}. \tag{2.65}$$

Note that the 4D power spectrum can be constructed as

$$P_j = \sum_{m',m=-j}^{j} \left(c^j_{m'm}\right)^* c^j_{m'm}. \tag{2.66}$$

The 4D bispectrum is invariant with respect to rotations of four-dimensional space, which include three-dimensional rotations. However, there are additional rotations, associated with the third polar angle θ_0, which, in our case, represents the radial information. In order to eliminate the invariance with respect to the third polar angle, we modified the atomic density as follows:

$$\rho_i(\mathbf{r}) = \delta(\mathbf{0}) + \sum_j \delta(\mathbf{r} - \mathbf{r}_{ij}), \tag{2.67}$$

i.e. by adding the central atom as a reference point.

The magnitude of the elements of the bispectrum scale as the cube of the number of neighbours, so we take the cube-root of the coefficients in order to make the comparison of different spectra easier.

2.3.5 *Results*

In practice, the infinite spherical harmonic expansion of the atomic neighbourhood is truncated to obtain a finite array of bispectral invariants. In Fig. 2.5 we show the 4D bispectra of atoms in a variety of environments, truncated to $j \leq 4$, which gives 42 bispectrum coefficients. In each case the r_0 parameter was set to highlight differences between the bispectral elements.

It can be seen from Fig. 2.5 that the bispectrum is capable of distinguishing very subtle differences in atomic neighbourhood environments. Some points of particular interest are the following. The difference between the face-centred cubic (fcc) and the hexagonal close-packed (hcp) structures is very small within the first neighbour shell, as is the difference between the corresponding bispectra (panel a). However, the difference is much more pronounced once second neighbours are included (panel b). The difference between the cubic and hexagonal diamond lattices is the stacking order of the (111) sheets. The positions of the four nearest neighbours and nine atoms of the second-nearest neighbour shell are the same and, only the positions of the remaining three neighbours are different, as shown in Fig. 2.6. The curves in Fig. 2.5c reflect the similarity of these two structures: most of the bispectrum coefficients are equal, except a few, which can be used for distinguishing the structures. Figure 2.5d shows the bispectra of three atoms in perfect diamond lattices, which differ in the lattice constants. This plot illustrates the sensitivity of the bispectrum in the radial dimension because the expansion of a lattice leaves all angular coordinates the same. It can be seen that the first element of the bispectrum array remains the same, because this is proportional only to the number of neighbours.

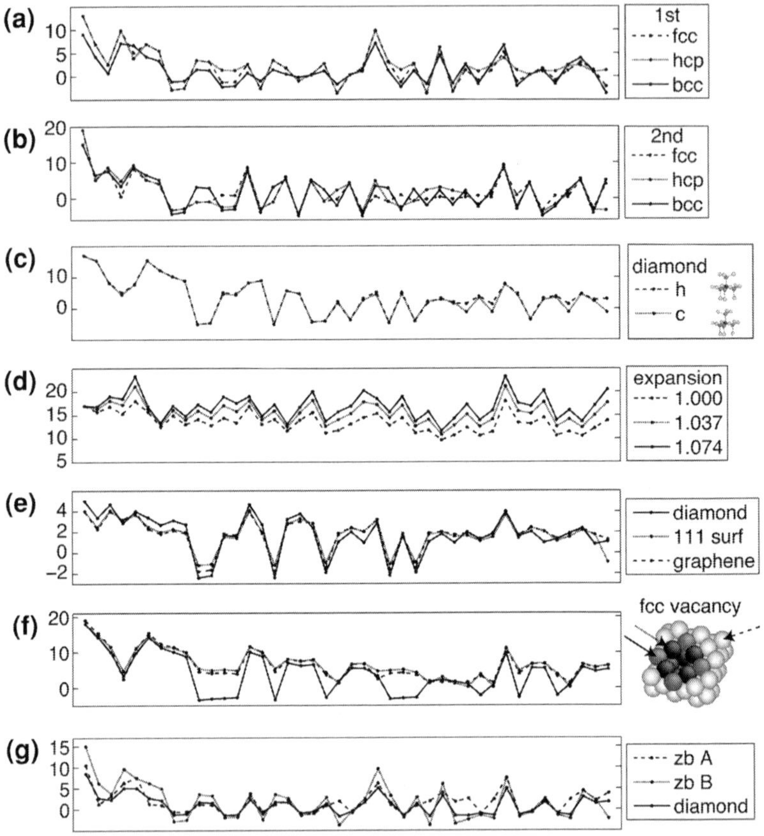

Fig. 2.5 Four-dimensional bispectra of atoms in various structures: **a** fcc/hcp/bcc lattices with a first neighbour cutoff; **b** fcc/hcp/bcc lattices with a second neighbour cutoff; **c** hexagonal and cubic diamond lattice; **d** expansion of a diamond lattice; **e** bulk diamond, (111) surface of diamond and graphene; **f** fcc vacancy; **g** the *A* and *B* atoms in a zincblende structure, compared with diamond

Fig. 2.6 Cubic and hexagonal diamond. Cubic diamond is shown in the *left*

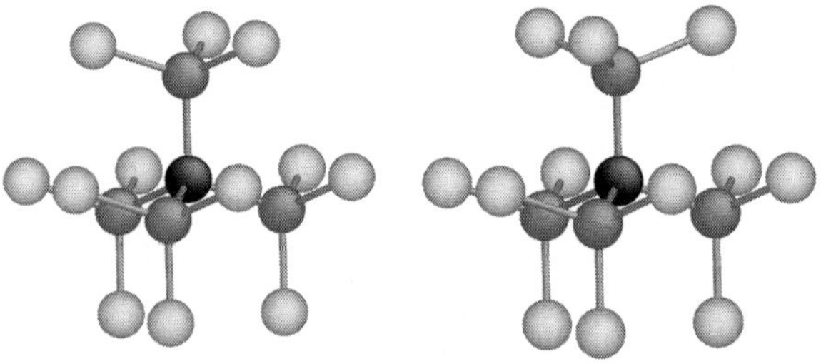

We performed the principle component analysis [16] on the bispectra of atoms in a slab of silicon. On the surface of the slab, the atoms were arranged according to the 7×7 reconstruction [17]. The position of the atoms were randomised by

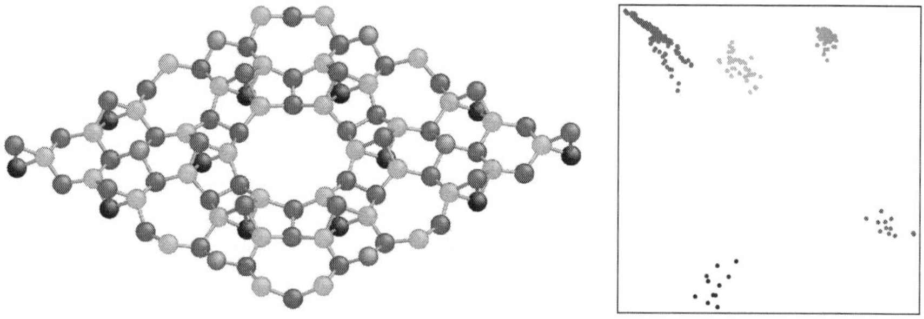

Fig. 2.7 Principle component analysis of the bispectrum of atoms on the 7 × 7 reconstruction of the (111) surface of silicon

0.3 Å. We projected the 42-dimensional space of the bispectrum—which corresponds to $j \leq 4$—to the two-dimensional plane and clustered the points using the k-means algorithm [18]. In Fig. 2.7, we show the result of the principle component analysis. Different colours are assigned to each cluster identified by the k-means method, and we coloured the atoms with respect to the cluster they belong. This example demonstrates that the bispectrum can be used to identify atomic environments in an automatic way.

It is straightforward to describe multi-species atomic environments using the bispectrum. We modify the atomic density function defined in Eq. 2.18 as

$$\rho_i(\mathbf{r}) = s_i\delta(\mathbf{0}) + \sum_j s_j\delta(\mathbf{r} - \mathbf{r}_{ij}), \tag{2.68}$$

where $\mathbf{s}$ contains an arbitrary set of coefficients, different for each species, which are thus distinguished. Figure 2.5g shows the resulting bispectra for the two different atoms in the zincblende lattice, as well as the diamond lattice for comparison. It can be seen that the bispectrum successfully distinguishes between the different species.

References

1. C.J. Pickard, R.J. Needs, Nat. Mater. **7**, 775 (2008)
2. D.J. Wales, *Energy Landscapes*. (Cambridge University Press, Cambridge, 2003)
3. J. Behler, M. Parrinello, Phys. Rev. Lett. **98**, 146401 (2007)
4. P.J. Steinhardt, D.R. Nelson, M. Ronchetti, Phys. Rev. B **28**, 784 (1983)
5. J.S. van Duijneveldt, D. Frenkel, J. Chem. Phys. **96**, 4655 (1992)
6. E.R. Hernández, J. Íñiguez, Phys. Rev. Lett. **98**, 055501 (2007)
7. A. van Blaaderen, P. Wiltzius, Science **270**, 1177 (1995)
8. R. Kakarala, Triple Correlation on Groups, Ph.D. thesis, Department of Mathematics, UC Irvine, 1992
9. S.A. Dianat, R.M. Rao, Opt. Eng. **29**, 504 (1990)
10. H. Weyl, *The Theory of Groups and Quantum Mechanics*. (Methuen, London, 1931)

11. P. Borwein, T. Erdélyi, *Polynomials and Polynomial Inequalities*. (Springer, New York, 1995)
12. D. Moroni, P.R. ten Wolde, P.G. Bolhuis, Phys. Rev. Lett. **94**, 235703 (2005)
13. R. Kondor, http://arxiv.org/abs/cs.CV/0701127, 2007
14. C.D. Taylor, Phys. Rev. B **80**, 024104 (2009)
15. D.A. Varshalovich, A.N. Moskalev, V.K. Khersonskii, Quantum Theory of Angular Momentum (World Scientific,Teaneck, 1987)
16. K. Pearson, Phil. Mag. **2**, 559 (1901)
17. K.C. Pandey, Physica **117**, **118**, 761 (1983)
18. A.K. Jain, M.N. Murty, P.J. Flynn, ACM Comp. Surv. **31**, 264 (2000)

Chapter 3
Gaussian Process

3.1 Introduction

Regression methods are important tools in data analysis. Parametric models can be expressed in functional forms that contain free parameters that are fitted such that the models reproduce observations. The model can often be formulated in a way that the functional form is a linear combination of the parameters. The fitting procedure in such cases is called linear regression. Nonlinear regression is needed if the functional form cannot be expressed as a simple linear combination of the parameters, but this case does not differ conceptually from the linear case. However, there is often no theory or model describing a particular process—or it is just too complicated to write the model in a closed functional form—but it is still important to make predictions of the outcome of the process. Non-parametric approaches, such as neural networks or Gaussian Processes, can be used to approximate the underlying function given a set of previously collected data. As neural network methods form a subset of Gaussian Processes [1], we decided to use the latter approach in our work.

3.2 Function Inference

Gaussian Processes predict the values of a function whose form is not explicitly known by using function observations as evidence. If $\mathbf{t} = \{t_i\}_{i=1}^{N}$ are values of a function $f : \mathbb{R}^n \rightarrow \mathbb{R}$ measured at the points $\mathbf{X} = \{\mathbf{x}_i\}_{i=1}^{N}$ with some error, predicting the value t_{N+1} at $\mathbf{x}_{N+1}$ can be formulated as a Bayesian inference problem. Bayes' theorem states that

$$P(t_{N+1}|\mathbf{t}) = \frac{P(\mathbf{t}|t_{N+1})P(t_{N+1})}{P(\mathbf{t})} \propto P(\mathbf{t}|t_{N+1})P(t_{N+1}), \tag{3.1}$$

A. Bartók-Pártay, *The Gaussian Approximation Potential*, Springer Theses,
DOI: 10.1007/978-3-642-14067-9_3,

where $P(t_{N+1})$ is a Gaussian prior on the function space. It is possible to introduce a Gaussian prior on function f as

$$f(\mathbf{x}) = \sum_h w_h \phi_h(\mathbf{x}) \tag{3.2}$$

where $\{\phi_h\}_{h=1}^H$ form a complete basis set and the distribution of $\mathbf{w}$ is a Gaussian with zero mean and variance σ_h:$w_h \sim N(0, \sigma_h)$. Each function value f_n is a linear combination of the basis functions:

$$f_n = \sum_{h=1}^H w_h \phi_h(\mathbf{x}_n) = \sum_{h=1}^H w_h R_{nh}, \tag{3.3}$$

where $R_{nh} \equiv \phi_h(\mathbf{x}_n)$. The covariance matrix of the function values $\mathbf{f}$ is the matrix of expectation values

$$\mathbf{Q} = \langle \mathbf{f}\,\mathbf{f}^T \rangle = \langle \mathbf{R}\mathbf{w}\mathbf{w}^T\mathbf{R}^T \rangle = \mathbf{R}\langle \mathbf{w}\mathbf{w}^T \rangle \mathbf{R}^T = \sigma_h^2 \mathbf{R}\mathbf{R}^T. \tag{3.4}$$

Thus the prior distribution of $\mathbf{f}$ is $N(\mathbf{0}, \mathbf{Q}) = N(\mathbf{0}, \sigma_h^2\mathbf{R}\mathbf{R}^T)$. However, each measurement contains noise, which we assume to be Gaussian with zero mean and variance σ_v. The vector of data points also has Gaussian distribution: $P(\mathbf{t}) \sim N(0, \mathbf{Q} + \sigma_v^2)$. We denote the covariance matrix of $\mathbf{t}$ by $\mathbf{C} \equiv \mathbf{Q} + \sigma_v^2\mathbf{I}$.

The distribution of the joint probability of observing t_{N+1} having previously observed $\mathbf{t}$ can be written as

$$P(t_{N+1}|\mathbf{t}) \propto P([\mathbf{t}\, t_{N+1}]), \tag{3.5}$$

where $P([\mathbf{t}t_{N+1}]) \sim N(0, \mathbf{C}_{N+1})$, or explicitly

$$P([\mathbf{t}\, t_{N+1}]) \propto \exp\left(-\frac{1}{2}[\mathbf{t}\, t_{N+1}]^T C_{N+1}^{-1} [\mathbf{t}\, t_{N+1}]\right). \tag{3.6}$$

The covariance matrix $\mathbf{C}_{N+1}$ and its inverse can be written as

$$\mathbf{C}_{N+1} = \begin{bmatrix} \mathbf{C}_N & \mathbf{k} \\ \mathbf{k}^T & \kappa \end{bmatrix} \tag{3.7}$$

and

$$\mathbf{C}_{N+1}^{-1} = \begin{bmatrix} \mathbf{M} & \mathbf{m} \\ \mathbf{m}^T & m \end{bmatrix}. \tag{3.8}$$

The submatrices of $\mathbf{C}_{N+1}^{-1}$ can be calculated via

$$\mathbf{C}_{N+1}\mathbf{C}_{N+1}^{-1} = \begin{bmatrix} \mathbf{C}_N\mathbf{M} + \mathbf{k}\mathbf{m}^\top & \mathbf{C}_N\mathbf{m} + m\mathbf{k} \\ \mathbf{k}^\top\mathbf{M} + \kappa\mathbf{m}^\top & \mathbf{k}^\top\mathbf{m} + \kappa m \end{bmatrix} = \mathbf{I}, \tag{3.9}$$

which leads to

$$m = \left(\kappa - \mathbf{k}^T \mathbf{C}_N^{-1} \mathbf{k}\right)^{-1} \tag{3.10}$$

$$\mathbf{m} = -m\mathbf{C}_N^{-1}\mathbf{k} \tag{3.11}$$

$$\mathbf{M} = \mathbf{C}_N^{-1} + \frac{1}{m}\mathbf{m}\mathbf{m}^T. \tag{3.12}$$

Substituting these into Eq. 3.6, we obtain

$$P(t_{N+1}|\mathbf{t}) \propto \exp\left(-\frac{\left(t_{N+1} - \hat{t}_{N+1}\right)^2}{2\sigma^2_{\hat{t}_{N+1}}}\right), \tag{3.13}$$

where the new variables $\hat{t}_{N+1}$ and $\sigma_{\hat{t}_{N+1}}$ are defined as

$$\hat{t}_{N+1} \equiv \mathbf{k}^T \mathbf{C}_N^{-1}\mathbf{t} \tag{3.14}$$

and

$$\sigma^2_{\hat{t}_{N+1}} \equiv \kappa - \mathbf{k}^T \mathbf{C}_N^{-1}\mathbf{k}, \tag{3.15}$$

i.e. t_{N+1} has Gaussian distribution with mean $\hat{t}_{N+1}$ and variance $\sigma^2_{\hat{t}_{N+1}}$. We use this formula to predict function values and error bars.

Figure 3.1 shows a one-dimensional example of the Gaussian Process regression. We sampled an arbitrary function at ten random points between the interval $(\frac{1}{4}, \frac{3}{4})$ and used these samples as the training points. We present the predicted values and the predicted errors in the entire interval (0, 1). It can be seen that inside the fitting region, the predicted values are very close to the original functions, and the predicted variance is also small. Outside the fitting region, the prediction is meaningless, and this is indicated by the large variance.

3.2.1 Covariance Functions

The elements of the covariance matrix $\mathbf{Q}$ defined in Eq. 3.4 can be determined as

$$Q_{nn'} = \sigma_h^2 \sum_h R_{nh}R_{n'h} = \sigma_h^2 \sum_h \phi_h(\mathbf{x}_n)\phi_h(\mathbf{x}_{n'}). \tag{3.16}$$

In our work, we used Gaussians centred at different points as basis functions. In one dimension, these would have the form

$$\phi_h(x) = \exp\left(-\frac{(x - x_h)^2}{2r^2}\right). \tag{3.17}$$

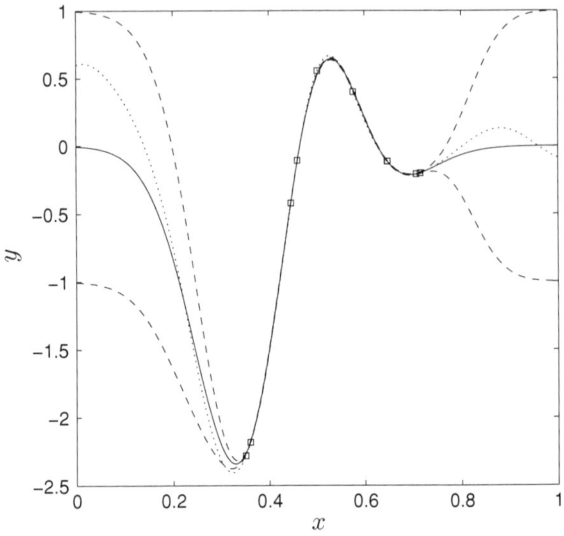

Fig. 3.1 Gaussian Process regression in one dimension. The original function (dotted line) was sampled at ten random points (open squares). The predicted function values (solid line) and the errors (dashed line) are shown

If the basis set consists of infinitely many basis functions which are distributed uniformly, the summation in Eq. 3.16 can be replaced by an integration:

$$Q_{nn'} \propto \int \exp\left(-\frac{(x_n - x_h)^2}{2r^2}\right) \exp\left(-\frac{(x_{n'} - x_h)^2}{2r^2}\right) \mathrm{d}r. \tag{3.18}$$

The integral of the product of two Gaussian is also a Gaussian, leading to the final expression—also known as the *kernel*—of the covariance matrix elements

$$Q_{nn'} = \delta^2 \exp\left(-\frac{(x_n - x_{n'})^2}{2\theta^2}\right), \tag{3.19}$$

where δ and θ are usually referred to as hyperparameters. This finding demonstrates that the Gaussian Process method is, in fact, an example of non-parametric regression with infinitely many basis functions, but where it is not necessary to determine the coefficients of the basis functions explicitly. We note that using Gaussians as basis functions is a convenient choice, as the elements of the covariance matrix can be calculated analytically using a simple Gaussian kernel, but depending on the nature of the target function, there is a large variety of alternative basis functions and kernels.

In the case of multidimensional input data, the Gaussian kernel could be modified such that different length scales are associated with different directions:

$$Q_{nn'} = \delta^2 \exp\left(-\frac{1}{2}\sum_i \frac{(x_{ni} - x_{n'i})^2}{\theta_i^2}\right), \tag{3.20}$$

where the vector $\boldsymbol{\theta} \equiv \{\theta_i\}_{i=1}^N$ contains the typical decorrelation length of the function in each dimension i. If we assume that the initial Gaussian basis functions are not aligned in the directions of the original input vectors, the kernel can be written in the form

$$Q_{nn'} = \delta^2 \exp\left(-\frac{1}{2} \mathbf{x}_n^T \boldsymbol{\Theta}^T \boldsymbol{\Theta} \mathbf{x}_n \right), \tag{3.21}$$

where $\boldsymbol{\Theta}$ is the matrix of hyperparameters.

3.2.2 Hyperparameters

The choice of hyperparameters δ, $\boldsymbol{\theta}$ and σ_v depends strongly on the data set. $\boldsymbol{\theta}$ represents the width of the basis functions, i.e. it characterises the typical length scale over which the function values become uncorrelated. δ places a prior on the variance of the parameter vector $\mathbf{w}$, describing the typical variance of the function, while σ_v is the assumed noise in the measured data values. Ideally, a prediction for t_{N+1} would be made by evaluating the integral

$$P(t_{N+1}|x_{N+1}, \mathbf{t}, \mathbf{X}) = \int P(t_{N+1}|x_{N+1}, \mathbf{t}, \mathbf{X}, \mathbf{h}) P(\mathbf{h}|\mathbf{t}, \mathbf{X}) d\mathbf{h}, \tag{3.22}$$

but depending on the model, the analytic form of the integral may or may not be known. Although it is always possible to carry out the integration numerically, for example, by Markov chain Monte Carlo or Nested Sampling [2], a computationally less demanding method is to approximate the integral at the most probable value of $\mathbf{h}$. It is often possible to choose good hyperparameters based on known features of the function, but the hyperparameters can also be optimised if needed. If we consider the probability distribution of a hyperparameter set $\mathbf{h}$ given a data set D:

$$P(\mathbf{h}|D) \propto P(D|\mathbf{h}) P(\mathbf{h}), \tag{3.23}$$

optimal hyperparameters can be obtained by maximising this probability, known as the marginal likelihood. Assuming a uniform prior on the hyperparameters and using the result found in Eq. 3.4, i.e. $P(\mathbf{t}|\mathbf{X}) \sim N(0, \mathbf{C})$, the logarithm of the likelihood is

$$\ln P(\mathbf{t}|\mathbf{X}, \mathbf{h}) = -\frac{1}{2} \mathbf{t}^T \mathbf{C}^{-1} \mathbf{t} - \frac{1}{2} \ln \det \mathbf{C} - \frac{N}{2} \ln 2\pi. \tag{3.24}$$

Maximising the logarithm of the likelihood with respect to the hyperparameters can be performed by gradient-based methods such as Conjugate Gradients [3], where that gradients can be calculated as

$$\frac{\partial \ln P}{\partial h_i} = \frac{1}{2}\mathbf{t}^T\mathbf{C}^{-1}\frac{\partial \mathbf{C}}{\partial h_i}\mathbf{C}^{-1}\mathbf{t} - \frac{1}{2}\mathrm{tr}\left(\mathbf{C}^{-1}\frac{\partial \mathbf{C}}{\partial h_i}\right). \tag{3.25}$$

3.2.3 Predicting Derivatives and Using Derivative Observations

Predicting the values of derivatives using a Gaussian Process can be performed by simply differentiating the expectation value $\hat{t}$ in Eq. 3.14:

$$\frac{\partial \hat{t}}{\partial x_i} = \frac{\partial \mathbf{k}^T}{\partial x_i}\mathbf{C}_N^{-1}\mathbf{t}. \tag{3.26}$$

The elements of $\mathbf{k}$ are given by the covariance function, hence we need to differentiate the covariance function,

$$\frac{\partial k_n}{\partial x_i} = \frac{\partial C(\mathbf{x}_n, \mathbf{x})}{\partial x_i} \tag{3.27}$$

which gives

$$\frac{\partial k_n}{\partial x_i} = \frac{x_{ni} - x_i}{\theta_i^2}\delta^2 \exp\left(-\frac{1}{2}\sum_i \frac{(x_{ni} - x_i)^2}{\theta_i^2}\right) \tag{3.28}$$

in the case of Gaussian kernels.

It is also possible that values of derivatives have been measured and these are also available. In order to use this data, we differentiate Eq. 3.2

$$\left.\frac{\partial f}{\partial x_i}\right|_{\mathbf{x}_n} = \sum_h w_h \left.\frac{\partial \phi_h}{\partial x_i}\right|_{\mathbf{x}_n}, \tag{3.29}$$

thus we need to substitute $\frac{\partial \phi_h}{\partial x_i}$ for the basis functions in Eq. 3.16 to give

$$Q_{nn'} = \sigma_h^2 \sum_h R_{nh}R_{n'h} = \sigma_h^2 \sum_h \left.\frac{\partial \phi_h}{\partial x_i}\right|_{\mathbf{x}_n} \phi_h(\mathbf{x}_{n'}). \tag{3.30}$$

$$Q_{nn'} = \sigma_h^2 \sum_h R_{nh}R_{n'h} = \sigma_h^2 \sum_h \left.\frac{\partial \phi_h}{\partial x_i}\right|_{\mathbf{x}_n} \left.\frac{\partial \phi_h}{\partial x_j}\right|_{\mathbf{x}_{n'}}. \tag{3.31}$$

For Gaussian kernels, the covariance between a derivative and a function value observation is

$$Q_{nn'} = \frac{x_{ni} - x_{n'i}}{\theta_i^2}\delta^2 \exp\left(-\frac{1}{2}\sum_k \frac{(x_{nk} - x_{n'k})^2}{\theta_i^2}\right), \tag{3.32}$$

or between two derivative observations the covariance is

$$Q_{nn'} = \left(\frac{1}{\theta_i \theta_j} - \frac{1}{2} \frac{x_{ni} - x_{n'i}}{\theta_i^2} \frac{x_{nj} - x_{n'j}}{\theta_j^2} \right) \delta^2 \exp\left(-\frac{1}{2} \sum_k \frac{(x_{nk} - x_{n'k})^2}{\theta_k^2} \right). \tag{3.33}$$

Finally, if the function is a composite function of the form $f(\mathbf{x}) \equiv f(\mathbf{y}(\mathbf{x}))$ and the derivatives $\frac{\partial f}{\partial x_i}$ are available, the Gaussian covariance function between a derivative (n-th) and function value (n'-th) observation is

$$Q_{nn'} = \sum_k \frac{y_{nk} - y_{n'k}}{\theta_k^2} \frac{\partial y_{nk}}{\partial x_i} \delta^2 \exp\left(-\frac{1}{2} \sum_k \frac{(y_{nk} - y_{n'k})^2}{\theta_k^2} \right), \tag{3.34}$$

and between two derivative observations $\frac{\partial f}{\partial x_i}$ and $\frac{\partial f}{\partial x_j}$ is

$$Q_{nn'} = \left(\sum_k \frac{1}{\theta_k^2} \frac{\partial y_{nk}}{\partial x_i} \frac{\partial y_{n'k}}{\partial x_j} - D_{ij} \right) \delta^2 \exp\left(-\frac{1}{2} \sum_k \frac{(y_{nk} - y_{n'k})^2}{\theta_k^2} \right), \tag{3.35}$$

with

$$D_{ij} = \frac{1}{2} \left(\sum_k \frac{y_{nk} - y_{n'k}}{\theta_k^2} \frac{\partial y_{nk}}{\partial x_i} \right) \left(\sum_k \frac{y_{nk} - y_{n'k}}{\theta_k^2} \frac{\partial y_{n'k}}{\partial x_j} \right). \tag{3.36}$$

Using the same model for observations of function values and their derivatives enables us to incorporate the available information into a single regression allowing us to infer both function values and derivatives.

Since there is no reason to assume that the noise is the same in case of both the function value and derivative observations, we use two distinct noise hyperparameters.

3.2.4 Linear Combination of Function Values

It is possible that linear combinations of function values can be observed during the data collection process:

$$f'_m = \sum_n L_{mn} f(\mathbf{x}_n) = \sum_{n,h} L_{mn} R_{nh} w_h. \tag{3.37}$$

If this is the case, Eq. 3.4 is thereby modified, so the covariance matrix of the observed values can be obtained as

$$\mathbf{Q}' = \langle \mathbf{f}'\mathbf{f}'T \rangle = \langle \mathbf{LRww}^T\mathbf{R}^T\mathbf{L}^T \rangle = \sigma_h^2 \mathbf{LRR}^T\mathbf{L}^T = \mathbf{LQL}^T. \tag{3.38}$$

In our work, Eq. 3.38 proved to be very useful, as only the total energy of an atomic system can be obtained using quantum mechanical calculations. However, we view the energy as arising from the sum of atomic contributions. Thus, in this case, the matrix $\mathbf{L}$ describing the relationship of the observations (total energy) to the unknown function values (atomic energies) consists of 0s and 1s.

3.2.5 Sparsification

Snelson and Ghahramani [4] introduced a modification to the standard Gaussian Process regression model for large, correlated data sets. The computational cost of the training process described in Eq. 3.13 scales as the cube of the number of data points, due to the computational cost of inverting the covariance matrix. In case of large data sets, the training process can become computationally expensive. Although the computational cost of predicting function values scales linearly with the number of teaching points, this cost can also be computationally demanding. If the data set is highly correlated, i.e. observations are made at closely spaced points, it is feasible to use a sparse approximation of the full Gaussian Process, which has significantly reduced computational requirements but only a little less accuracy.

We used the sparsification procedure described in [4]. In the sparsification procedure, a set of M pseudo-inputs $\{\bar{\mathbf{x}}_m\}_{m=1}^{M}$ are chosen from the full data set of N input values $\{\mathbf{x}_n\}_{n=1}^{N}$, and the covariance matrices $\mathbf{C}_{NM}$ and $\mathbf{C}_M$ are calculated as

$$[\mathbf{C}_M]_{mm'} = C(\bar{\mathbf{x}}_m, \bar{\mathbf{x}}_{m'}) \tag{3.39}$$

and

$$[\mathbf{C}_{NM}]_{nm} = [\mathbf{k}_n]_m = C(\mathbf{x}_n, \bar{\mathbf{x}}_m). \tag{3.40}$$

In order to simulate the full covariance matrix, the matrix

$$\mathbf{\Lambda} = \mathrm{Diag}(\mathrm{diag}(\mathbf{C}_N - \mathbf{C}_{NM}\mathbf{C}_M^{-1}\mathbf{C}_{MN})) \tag{3.41}$$

is also needed, where $\mathbf{C}_N$ is the full $N \times N$ covariance matrix, although only the diagonal elements are calculated. The elements of the covariance vector $\mathbf{k}$ are calculated from the coordinates of the pseudo-inputs and the test point $\mathbf{x}_*$:

$$k_m = C(\bar{\mathbf{x}}_m, \mathbf{x}_*). \tag{3.42}$$

The pseudo-covariance matrix of the sparsified data set is

$$\mathbf{Q}_M = \mathbf{C}_M + \mathbf{C}_{MN}(\mathbf{\Lambda} + \sigma^2\mathbf{I})^{-1}\mathbf{C}_{NM}, \tag{3.43}$$

which can now be used to predict the function value and the error estimate at the test point as

$$\hat{t} = \mathbf{k}^T\mathbf{Q}_M^{-1}\mathbf{C}_{MN}(\mathbf{\Lambda} + \sigma^2\mathbf{I})^{-1}\mathbf{t} \tag{3.44}$$

$$\sigma_{\hat{t}}^2 = C(\mathbf{x}_*, \mathbf{x}_*) - \mathbf{k}_*^T(\mathbf{C}_M^{-1} - \mathbf{Q}_M^{-1})\mathbf{k} + \sigma^2. \tag{3.45}$$

In order to obtain an optimal set of hyperparameters and pseudo-inputs, the likelihood function

$$\begin{aligned} \log L = &-\frac{1}{2}\mathbf{t}^T(\mathbf{C}_{NM}\mathbf{C}_M^{-1}\mathbf{C}_{MN} + \mathbf{\Lambda} + \sigma^2\mathbf{I})^{-1}\mathbf{t} \\ &-\frac{1}{2}\log|\mathbf{C}_{NM}\mathbf{C}_M^{-1}\mathbf{C}_{MN} + \mathbf{\Lambda} + \sigma^2\mathbf{I}| - \frac{n}{2}\log 2\pi \end{aligned} \tag{3.46}$$

is maximised in the space of hyperparameters and pseudo-inputs.

In our work, observation of single function values is not possible, i.e. only total energies (sum of atomic energies) and forces (sum of derivatives of local energies) are accessible. Depending on the number of atoms in the cell, in the case of total energy observations, and the number of atoms within the chosen cutoff radius, in the case of force observations, a large number of input values has to be added to the training set, regardless of whether the neighbourhood of a particular atom is different from the ones previously encountered. Thus in our case, the sparsification process is crucial in order to develop a tractable computational scheme.

References

1. C.E. Rasmussen, C.K.I. Williams, *Gaussian Processes for Machine Learning* (MIT Press, Cambridge, MA, USA, 2006)
2. J. Skilling, J. Bayesian Anal. **1**, 833 (2006)
3. M.R. Hestenes, E. Stiefel, J. Res. Natl. Bur. Stand. **49**, 409 (1952)
4. E. Snelson, Z. Ghahramani, in *Sparse Gaussian Processes Using Pseudo-inputs*, ed. by Y. Weiss, B. Schölkopf, J. Platt. Advances in Neural Information Processing Systems 18 (MIT Press, Cambridge, MA, 2006), pp. 1257–1264

Chapter 4
Interatomic Potentials

4.1 Introduction

A wide variety of models have been developed to describe atomic interactions, ranging from the very accurate and extremely expensive to the fast but very approximate. Quantum Mechanics ultimately provides a true description of matter via solving the Schrödinger equation, but even in its crudest approximation, the use of Quantum Mechanics is limited to a few hundreds of atoms or a few hundreds of different configurations, which is inadequate to sample the entire phase space of a system. A series of further simplifications leads to the realm of analytic potentials that can be used to describe larger systems or more configurations. The so-called empirical potentials are based on fixed functional forms, which are equally based on theoretical considerations and intuition, making the creation of new potentials a combination of "art and science" [1]. Analytic potentials can be described as non-linear parametric regression from the statistical point of view, where the fitting process is based on experimental or quantum mechanical data. Further, the parametric formula that is chosen to describe the behaviour of the real system is often fitted to reproduce some well-known equilibrium properties, such as the lattice constant and elastic constants of the bulk material or the structure of a liquid, and it is assumed that the same function will perform well in very different configurations. This clearly implies that analytic potentials are expected to be able to extrapolate to very different environments on the basis of the physical insight used when the particular functional form was chosen. Even if there exists such a functional form, it follows from the overly complicated nature of regression in such high dimensions that finding the right form and fitting it to each new interesting material is extremely difficult. Our work focuses on the development of a potential based on non-linear, non-parametric regression methods that infers the interactions directly from quantum mechanical data, though the approach can be adopted irrespective of the origin of the data.

A. Bartók-Pártay, *The Gaussian Approximation Potential*, Springer Theses,
DOI: 10.1007/978-3-642-14067-9_4, © Springer-Verlag Berlin Heidelberg 2010

4.2 Quantum Mechanics

In the general case, the Schrödinger equation takes the form

$$i\hbar\frac{\partial\Psi(\mathbf{r},t)}{\partial t} = \hat{H}\Psi(\mathbf{r},t), \tag{4.1}$$

where Ψ is the time-dependent wave-function, $\mathbf{r}$ contains the coordinates of all the particles in the system and $\hat{H}$ is the Hamiltonian operator. The Hamiltonian can be written as

$$\hat{H} = -\sum_i \frac{\hbar^2}{2m_i}\nabla_i^2 + V(\mathbf{r}), \tag{4.2}$$

where $V(\mathbf{r})$ is the potential energy. The standing wave solution of the time dependent Schrödinger equation is

$$\Psi(\mathbf{r},t) = \psi(\mathbf{r})\exp\left(-\frac{iEt}{\hbar}\right), \tag{4.3}$$

which leads to the time-independent form of the Schrödinger equation

$$\hat{H}\psi(\mathbf{r}) = E\psi(\mathbf{r}). \tag{4.4}$$

Atomic systems consist of electrons and nuclei, hence Eq. 4.2 becomes

$$\begin{aligned}\hat{H} = &-\sum_i^{\text{elec.}} \frac{\hbar^2}{2m_e}\nabla_i^2 + \sum_{i<j}^{\text{elec.}} \frac{q_e^2}{r_{ij}} - \sum_i^{\text{elec.}}\sum_A^{\text{nuclei}} Z_A\frac{q_e^2}{r_{iA}} \\ &-\sum_A^{\text{nuclei}} \frac{\hbar^2}{2m_A}\nabla_A^2 + \sum_{A<B}^{\text{nuclei}} Z_A Z_B\frac{q_e^2}{r_{AB}}\end{aligned} \tag{4.5}$$

where m_e and q_e are the mass and the charge of an electron, m_A and Z_A are the mass and atomic number of the nucleus A. The Born–Oppenheimer approximation further simplifies the solution of Eq. (4.4) by assuming that the coupling of the electrons and nuclei is negligible. The basis of this assumption is that the mass of the nuclei is at least three order of magnitudes larger than the mass of the electrons, thus the electrons adapt to the nuclei adiabatically. The Born–Oppenheimer approximation can be expressed as

$$\begin{aligned}&\left(-\sum_i^{\text{elec.}} \frac{\hbar^2}{2m_e}\nabla_i^2 + \sum_{i<j}^{\text{elec.}} \frac{q_e^2}{r_{ij}} - \sum_i^{\text{elec.}}\sum_A^{\text{nuclei}} Z_A\frac{q_e^2}{r_{iA}}\right)\psi(\mathbf{r},\mathbf{R}) \\ &+\sum_{A<B}^{\text{nuclei}} Z_A Z_B\frac{q_e^2}{r_{AB}} = E_e(\mathbf{R})\psi(\mathbf{r},\mathbf{R})\end{aligned} \tag{4.6}$$

$$\left(-\sum_{A}^{\text{nuclei}}\frac{\hbar^2}{2m_A}\nabla_A^2 + E_e(\mathbf{R})\right)\chi(\mathbf{R}) = E\chi(\mathbf{R}), \tag{4.7}$$

where the electronic wavefunction $\psi(\mathbf{r}, \mathbf{R})$ only depends on the coordinates of the electrons $\mathbf{r}$ and the coordinates of the nuclei $\mathbf{R}$ are regarded as parameters. The solutions of Eq. 4.6, the so-called electronic Schrödinger equation provides the *potential energy surface* (PES) $E_e(\mathbf{R})$, which describes the interactions of the nuclei. The nuclear Schrödinger equation is often replaced by the classical equations of motion.

4.2.1 Density Functional Theory

The analytic solution of the electronic Schrödinger equation is impossible for systems more complicated than a hydrogen molecular-ion H_2^+. There exists a wide range of methods that are concerned with determining the electronic structure, ranging from the very approximate tight-binding [2] approach to the essentially exact full configuration interaction [3] method. In our work, we used Density Functional Theory as the underlying quantum mechanical method.

Density Functional Theory aims to find the ground state electron density rather than the wavefunction.

$$\rho(\mathbf{r}) = \int |\psi(\mathbf{r}, \mathbf{r}_2, \ldots, \mathbf{r}_N)|^2 d\mathbf{r}_2, \ldots, d\mathbf{r}_N \tag{4.8}$$

The density depends only on three spatial coordinates instead of 3*N*, reducing the complexity of the task enormously. The Hohenberg–Kohn principles prove that the electron density is the most central quantity determining the electronic interactions and forms the basis of an exact expression of the electronic ground state.

4.2.1.1 The Hohenberg–Kohn principles

The basic lemma of Hohenberg and Kohn [4] states that the ground state electron density of a system of interacting electrons in an arbitrary external potential determines this potential uniquely. The proof is given by the variational principle. If we consider a Hamiltonian $\hat{H}_1$ of an external potential V_1 as

$$\hat{H}_1 = \hat{T} + \hat{U} + \hat{V}_1, \tag{4.9}$$

where $\hat{T}$ is the kinetic energy operator and $\hat{U}$ is the electron–electron interaction operator. The solution of the Schrödinger equation

$$\hat{H}_1\psi = E\psi \tag{4.10}$$

is the ground state wavefunction ψ_1, which corresponds to the electron density ρ_1. The ground state energy is then

$$E_1 = \langle\psi_1|\hat{H}_1|\psi_1\rangle = \int V_1(\mathbf{r})\rho(\mathbf{r}) + \langle\psi_1|\hat{T} + \hat{U}|\psi_1\rangle. \tag{4.11}$$

Considering another potential V_2, which cannot be obtained as V_1 + constant, with a ground state wavefunction ψ_2, which generates the same electron density, the ground state energy is

$$E_2 = \int V_2(\mathbf{r})\rho(\mathbf{r}) + \langle\psi_2|\hat{T} + \hat{U}|\psi_2\rangle. \tag{4.12}$$

According to the variational principle,

$$\begin{aligned} E_1 < \langle\psi_2|\hat{H}_1|\psi_2\rangle &= \int V_1(\mathbf{r})\rho(\mathbf{r}) + \langle\Psi_2|\hat{T} + \hat{U}|\Psi_2\rangle \\ &= E_2 + \int [V_1(\mathbf{r}) - V_2(\mathbf{r})]\rho(\mathbf{r}) \end{aligned} \tag{4.13}$$

and

$$\begin{aligned} E_2 < \langle\psi_1|\hat{H}_2|\psi_1\rangle &= \int V_2(\mathbf{r})\rho(\mathbf{r}) + \langle\psi_1|\hat{T} + \hat{U}|\psi_1\rangle \\ &= E_1 + \int [V_2(\mathbf{r}) - V_1(\mathbf{r})]\rho(\mathbf{r}). \end{aligned} \tag{4.14}$$

By adding the two inequalities together, we find the contradiction

$$E_1 + E_2 < E_1 + E_2. \tag{4.15}$$

This is the indirect proof that no two different external potentials can generate the same electron density.

The second Hohenberg–Kohn theorem establishes a link between the total energy and the electron density, namely that there exists a universal energy functional, which is valid for every external potential, and its global minimum corresponds to the ground state of the system and the ground state electron density. To prove this theorem, we write the total energy functional as a universal functional

$$E[\rho] = F_{\mathrm{HK}}[\rho] + \int V(\mathbf{r})\rho(\mathbf{r}) + E_{ZZ}, \tag{4.16}$$

where F_{HK} applies to every electronic system. It determines the entire electronic energy except the energy due to the external potential $V(\mathbf{r})$. E_{ZZ} is the interaction between the nuclei. The ground state energy is given by

$$E = \langle\psi|\hat{H}|\psi\rangle = E[\rho]. \tag{4.17}$$

According to the variational principle, changing the wavefunction to a different ψ', which in turn corresponds to a different electron density ρ', the resulting energy

$$E < E' = \langle\psi'|\hat{H}|\psi'\rangle, \tag{4.18}$$

is greater than E, thus ρ cannot correspond to the exact ground state.

We note that the ground state wavefunction can be found from the variational principle

$$E = \min_{\tilde{\psi}}\langle\tilde{\psi}|\hat{T} + \hat{U} + \hat{V}|\tilde{\psi}\rangle, \tag{4.19}$$

where $\tilde{\psi}$ is a trial wavefunction. The variational principle can be reformulated in terms of trial densities, $\tilde{\rho}$:

$$E = \min_{\tilde{\rho}} E[\tilde{\rho}] \tag{4.20}$$

4.2.1.2 The self-consistent Kohn–Sham equations

The Hohenberg–Kohn principles provide the theoretical basis of Density Functional Theory, specifically that the total energy of a quantum mechanical system is determined by the electron density through the Kohn–Sham functional. In order to make use of this very important theoretical finding, Kohn–Sham equations are derived, and these can be used to determine the electronic ground state of atomic systems.

The total energy of a system of interacting electrons in the external potential of the classic nuclei can be written as

$$E[\rho] = T[\rho] + E_{\mathrm{H}}[\rho] + E_{\mathrm{xc}}[\rho] + E_{Ze}[\rho] + E_{ZZ}, \tag{4.21}$$

where $T[\rho]$ is the kinetic energy functional, E_{xc} is the exchange-correlation functional, E_{H} is the Hartree interaction between electrons, E_{Ze} is the interaction between the electrons and the nuclei and E_{ZZ} is the nuclei–nuclei interaction. The latter three energies have the forms

$$E_{\mathrm{H}}[\rho] = \int \frac{\rho(\mathbf{r})\rho(\mathbf{r}')}{|\mathbf{r}-\mathbf{r}'|}\mathbf{d}\mathbf{r}\mathbf{d}\mathbf{r}' \tag{4.22}$$

$$E_{Ze}[\rho] = \sum_{A}^{\text{nuclei}} \int Z_A \frac{\rho(\mathbf{r})}{|\mathbf{r}-\mathbf{r}_A|}\mathbf{d}\mathbf{r} \tag{4.23}$$

$$E_{ZZ} = \sum_{A<B}^{\text{nuclei}} \frac{Z_A Z_B}{|\mathbf{r}_A - \mathbf{r}_B|}, \tag{4.24}$$

whereas the exact form of functionals $T[\rho]$ and $E_{\mathrm{xc}}[\rho]$ is not specified by the theory. However, according to the Hohenberg–Kohn principle, any system of interacting electrons can be described as a system of independent electrons moving in an effective potential, meaning that the kinetic energy functional can represented by the kinetic energy of non-interacting electrons, T_S. The difference between the true kinetic energy functional and T_S

$$\Delta T = T[\rho] - T_S \tag{4.25}$$

is included in the exchange-correlation functional, which still needs to be determined. The non-interacting kinetic energy operator T_S is simply written as

$$T_S = -\frac{\hbar^2}{2m_e}\sum_n^{\text{elec.}} \langle \psi_n | \nabla_n^2 | \psi_n \rangle, \tag{4.26}$$

where ψ_n are the independent electron orbitals. The one-electron orbitals determine the charge density as

$$\rho(\mathbf{r}) = \sum_n \psi_n^*(\mathbf{r})\psi_n(\mathbf{r}). \tag{4.27}$$

Hence the ground state will correspond to the electronic density at which the functional derivative of the total energy with respect to ψ_n is zero, while maintaining the orthogonality constraints

$$\langle \psi_i | \psi_j \rangle = \delta_{ij} \tag{4.28}$$

via the Lagrange multipliers $\epsilon_i j_{ij}$. Thus minimising the energy functional and the constraints

$$\frac{\delta\left[E - \sum_{ij} \epsilon_{ij}(\langle \psi_i | \psi_j \rangle - \delta_{ij})\right]}{\delta\psi_n} = 0, \quad \forall n \tag{4.29}$$

leads to the Kohn–Sham equations,

$$\begin{aligned} 0 &= \frac{\delta T_S}{\delta\psi_n} + \frac{E_H[\rho] + E_{xc}[\rho] + E_{Ze}[\rho]}{\delta\rho}\frac{\delta\rho}{\delta\psi_n} - \frac{\sum_{ij}\epsilon_{ij}\langle \psi_i | \psi_j \rangle}{\delta\psi_n} \\ &= \Delta_n\psi_n + \hat{V}_{\text{eff}}\psi_n - \sum_j \epsilon_{nj}\psi_j, \end{aligned} \tag{4.30}$$

which can be solved as n independent equations,

$$\Delta_n\psi_n' + \hat{V}_{\text{eff}}\psi_n' = \epsilon_n'\psi_n', \tag{4.31}$$

since there exists a basis set where the energy matrix is diagonal. Although the minimisation can be performed directly, as implemented in CASTEP as conjugate gradients for insulating systems or EDFT [5], an iterative approach is more often used. The effective potential $\hat{V}_{\text{eff}}$ depends on the electronic density, thus it is calculated using some initial guess for the density, then the Kohn–Sham equations are solved, resulting in a new density. This process is repeated until the electron density becomes self-consistent.

4.3 Empirical Potentials

The Born–Oppenheimer approximation, as given in Eq. 4.6, suggests that when considering solely the interactions between the nuclei, the electrons do not have to be explicitly taken in account. The reason why the Schrödinger equation has to be solved in many applications is the need for the accurate description of the Potential Energy Surface provided by Quantum Mechanics. If there were an alternative way to determine the Potential Energy Surface felt by the nuclei $V(\mathbf{R}) \equiv E(\mathbf{R})$, Quantum Mechanics could be bypassed entirely. Empirical potentials, as well as our research, aim to achieve this.

4.3.1 Hard-sphere Potential

The simplest interatomic potential is the hard-sphere potential, that can be characterised as

$$V(r) = \begin{cases} 0 & \text{if} \quad r \leq r_0 \\ \infty & \text{if} \quad r > r_0 \end{cases}, \tag{4.32}$$

where r_0 is the radius of the sphere. Even this simple functional form can describe the fact that atoms repel each other due to the Pauli exclusion principle, albeit in a rather crude way. As this potential completely lacks attractive terms, its use is usually limited to bulk phases. The hard-sphere model is often used for testing purposes, as despite of its simplicity, a system of hard-spheres shows a fluid–solid phase transition [6, 7]. More recently, systems of colloid particles were also modelled as hard spheres [8, 9], and the results of these simulations have received strong experimental support.

4.3.2 Lennard–Jones Potential

The Lennard–Jones Potential

$$V(r) = 4\epsilon\left(\frac{\sigma^{12}}{r^{12}} - \frac{\sigma^6}{r^6}\right) \tag{4.33}$$

was originally introduced to describe the interaction between argon atoms [10]. The two terms in the expression are the repulsion due to Pauli exclusion and the attraction which arises from dispersion interactions. The r^{-6} variation is obtained by considering the interaction of two induced dipoles on closed-shell atoms. Although the r^{-12} term has been introduced primarily because it is the square of the other term—therefore its computation is very efficient— and has no theoretical justification, the Lennard–Jones potential reproduces the properties of argon remarkably well [11]. In the case of other noble gases, quantum effects (for He and Ne), contribution from the interaction of higher order moments and relativistic effects (for Kr, Xe, Ra) become more significant and so the Lennard–Jones model is not so successful. The Lennard–Jones potential has been applied to different types of systems, because of the ease of computation and the strong physical basis. Potentials for ions are often built as Lennard–Jones spheres and point charges [12, 13], the most successful water models are based on partial charges and Lennard–Jones term(s) [14, 15], or even groups of atoms, such as methyl groups are modelled as a single Lennard–Jones particle [16]. While being a relatively simple potential, systems composed of Lennard–Jones particles show complex phase behaviour, which makes the use of this potential attractive as test systems in such studies and method development [17–19].

4.3.3 The Embedded-Atom Model

The embedded-atom model was developed by Daw and Baskes [20] and was originally intended to describe metallic systems. In general, the potential takes the form

$$E = \sum_i F(\rho_i) + \frac{1}{2}\sum_{j \neq i} \Phi(r_{ij}), \tag{4.34}$$

where ρ_i is the electron density at the centre of atom i due to the atoms at neighbouring sites

$$\rho_i = \sum_{j \neq i} \rho_j(r_{ij}), \tag{4.35}$$

where ρ_j is the electronic density of atom j. F is the embedding functional and Φ represents the core–core repulsion. This potential is derived from density functional theory, where the electron density is approximated by a sum of atomic contributions and the energy functional is substituted by a simple analytic function. The parameters in the embedded atom potentials used in the original applications were fitted to experimental observables, such as lattice constants and elastic moduli.

More recently, a particularly interesting new formulation of the embedded-atom model, called the force-matching method has been published by Ercolessi and Adams [21]. In this work, no prior assumptions were made on the actual functional forms in Eqs. 4.34 and 4.35. All functions were described by splines, and the splines were fitted such that the difference between the forces predicted by the model and the forces determined by first-principle calculations is minimal. This method is an early example of using a flexible regression for building interatomic potentials. The differences between the forces predicted by the Ercolessi–Adams potential and Density Functional Theory are remarkably small in bulk fcc aluminium, although the description of surfaces is less accurate.

4.3.4 The Modified Embedded-Atom Model

Although the embedded atom model proved to be a good potential for metallic systems, it fails to describe covalent materials, such as semiconductors. The reason for this is that the electron density in Eq. 4.35 is assumed to be isotropic, which is a good approximation in close packed systems, like fcc crystals, but in the case of covalent bonds, the electron density is higher along the bonds. In order to correct this, an angle-dependent density term was introduced by Baskes [22] for silicon

$$\rho_i = \sum_{j \neq i} \rho(r_{ij}) + \sum_{j \neq i, k \neq i} \rho(r_{ij})\rho(r_{ik})g(\cos\theta_{jik}), \tag{4.36}$$

where θ_{jik} is the bond angle between the *ji* and *ki* bonds. The original formulation used the fixed functional form

$$g(\cos\theta_{jik}) = 1 - 3\cos^2\theta_{jik} \tag{4.37}$$

for the angle-dependency, which biased the equilibrium bond angle preference to tetrahedral angles, resulting in a poor description of liquid or non-tetrahedral phases of silicon. Lenosky et al. [23] adopted the force-matching method for the modified embedded-atom model.

Taylor showed an elegant generalisation of the modified embedded atom model in [24]. In this work, he formulated a Taylor-expansion of the total energy functional around the ground-state density of atoms in terms of density variations, which led to a general expression for the total energy of the system as a function of the atomic coordinates. The energy of an atomic system is determined as a functional of the atomic density as

$$E = \Phi[\rho(\mathbf{r})], \tag{4.38}$$

where

$$\rho(\mathbf{r}) = \sum_i \delta(\mathbf{r} - \mathbf{r}_i) \tag{4.39}$$

and δ is the Dirac-delta function. This form is, in fact, an alternative description of the total energy as given by Density Functional Theory. The atomic density determines, through Poisson's equation, the external potential through which the electrons move as

$$\nabla^2 V_{\text{ext}} = -\frac{\rho}{\epsilon_0}, \tag{4.40}$$

which in turn corresponds to a ground state electron density and a total energy. If E_0 is the minimum of the total energy with respect to the atomic density, the energy can be expressed in a Taylor series in variations in the density $\rho = \rho_0 + \delta\rho$ as

$$E = E_0 + \int \left.\frac{\delta E}{\delta \rho}\right|_{\mathbf{r}} \delta\rho(\mathbf{r})\mathrm{d}\mathbf{r} + \iint \left.\frac{\delta^2 E}{\delta \rho^2}\right|_{\mathbf{r},\mathbf{r}'} \delta\rho(\mathbf{r})\delta\rho(\mathbf{r}')\mathrm{d}\mathbf{r}\mathrm{d}\mathbf{r}' + \cdots. \tag{4.41}$$

The density variation $\delta\rho$ is given by

$$\delta\rho(\mathbf{r}) = \sum_i \left[\delta(\mathbf{r} - \mathbf{r}_i) - \delta(\mathbf{r} - \mathbf{r}_i^0)\right], \tag{4.42}$$

where $\mathbf{r}_i^0$ are the equilibrium positions of the atoms, corresponding to the ground state atomic density. The first-order term in Eq. 4.41 disappears because the Taylor-expansion is performed around the minimum. Substituting 4.41 in Eq. 4.42, then integrating results in

$$E = E_0 = \sum_{i,j} \left.\frac{\delta^2 \Phi}{\delta \rho^2}\right|_{\mathbf{r}_i,\mathbf{r}_j} - \left.\frac{\delta^2 \Phi}{\delta \rho^2}\right|_{\mathbf{r}_i^0,\mathbf{r}_j} - \left.\frac{\delta^2 \Phi}{\delta \rho^2}\right|_{\mathbf{r}_i,\mathbf{r}_j^0} + \left.\frac{\delta^2 \Phi}{\delta \rho^2}\right|_{\mathbf{r}_i^0,\mathbf{r}_j^0}. \tag{4.43}$$

Introducing the new functions

$$f(\mathbf{r}_i, \mathbf{r}_j) = \left.\frac{\delta^2 \Phi}{\delta \rho^2}\right|_{\mathbf{r}_i,\mathbf{r}_j} \tag{4.44}$$

and

$$g(\mathbf{r}_i) = \sum_j \left.\frac{\delta^2 \Phi}{\delta \rho^2}\right|_{\mathbf{r}_i,\mathbf{r}_j^0} = \sum_j \left.\frac{\delta^2 \Phi}{\delta \rho^2}\right|_{\mathbf{r}_j^0,\mathbf{r}_i} \tag{4.45}$$

we can write the total energy as a sum of one- and two-body terms

$$E = E_0' + \sum_i g(\mathbf{r}_i) + \sum_{i,j} f(\mathbf{r}_i, \mathbf{r}_j) + \cdots. \tag{4.46}$$

Similarly, if we consider the local atomic densities around atom i

$$\delta_i(\mathbf{r}) = \sum_{j \neq i} \delta(\mathbf{r} - \mathbf{r}_j) w(r_{ij}), \tag{4.47}$$

where w is a screening function, we obtain the total energy expression up to second order

$$E = \sum_i E_{i,0} + \sum_i \sum_{j \neq i} g(\mathbf{r}_{ij}) w(r_{ij}) + \sum_i \sum_{j \neq i} \sum_{k \neq i} f(\mathbf{r}_{ij}, \mathbf{r}_{ik}) w(r_{ij}) w(r_{ik}). \tag{4.48}$$

This expression has the same form as the modified embedded atom model. Taylor represented the local atomic density by bond-order parameters and different radial functions as discussed in Sect. 2.3.1 in Chap. 2. By choosing appropriate radial functions, he obtained the original modified embedded-atom formula, but systematic improvement of the formula is also possible in his framework.

4.3.5 Tersoff Potential

The form of interatomic potential suggested by Tersoff [25] is an example of the wider family of bond-order potentials [26]. The total energy is written as a sum of pair like terms,

$$E = \frac{1}{2} \sum_{i \neq j} V_{ij} \tag{4.49}$$

$$V_{ij} = f_{\mathrm{cut}}(r_{ij}) [f_{\mathrm{R}}(r_{ij}) + b_{ij} f_{\mathrm{A}}(r_{ij})] \tag{4.50}$$

where f_{R} and f_{A} are repulsive and attractive terms, f_{cut} is a cutoff function, and b_{ij} is the bond-order term

$$f_{\mathrm{R}}(r_{ij}) = A_{ij} \exp(-\lambda_{ij} r_{ij}) \tag{4.51}$$

$$f_{\mathrm{A}}(r_{ij}) = -B_{ij} \exp(-\mu_{ij} r_{ij}) \tag{4.52}$$

$$f_{\mathrm{cut}}(r_{ij}) = \begin{cases} 1 & \text{if} \quad r_{ij} < R_{ij} \\ \frac{1}{2} + \frac{1}{2} \cos\left(\pi \frac{r_{ij} - R_{ij}}{S_{ij} - R_{ij}} \right) & \text{if} \quad R_{ij} < r_{ij} < S_{ij} \\ 0 & \text{if} \quad r_{ij} > S_{ij} \end{cases} \tag{4.53}$$

$$b_{ij} = \chi_{ij} \left(1 + \beta_i^{n_i} \zeta_{ij}^{n_i} \right)^{1/2n_i} \tag{4.54}$$

$$\zeta_{ij} = \sum_{k \neq i,j} f_{\mathrm{cut}}(r_{ik}) \omega_{ik} g(\theta_{ijk}) \tag{4.55}$$

$$g(\theta_{ijk}) = 1 + \frac{c_i^2}{d_i^2} - \frac{c_i^2}{d_i^2 + (h_i - \cos\theta_{ijk})^2}. \tag{4.56}$$

The resulting potential is, in fact, a many-body potential, as the bond-order terms depend on the local environment. Bond-order potentials can also be derived from a quantum mechanical method, tight-binding [26] and can be regarded as an analytical approximation of the solutions of the Schrödinger equation.

4.4 Long-range Interactions

The electrostatic contribution to the total energy is often not negligible. If there is charge transfer between atoms or polarisation effects are significant, the interaction between charges, dipoles or even higher order multipoles needs to be calculated. There are well-established methods to determine the electrostatic energy and forces, such as the Ewald-summation technique [27]. The central question is the values of the electric charges and multipoles in a particular model. In many cases fixed charges are used, for example, most water potentials [28] and models of ionic crystals [29] have predetermined charges. Classical water potentials describe the structure of bulk liquid water well, however, the representation of solutions is often poor due to the fact that these models no longer describe the interactions correctly in the modified environment and the resulting electric fields.

The electronegativity equalisation method [30] and the charge equilibration method [31] were designed to introduce charges which depend on the atomic environment and the local electric field. The atomic charges predicted by these methods agree well with the experimental values and with the ones determined by quantum mechanical methods for ionic crystals and organic molecules.

Electrostatic models including multipoles have also been developed. The multipoles are often deduced from the electronic structure determined by ab initio methods, for example, by using Wannier functions [32]. The dependence of the multipoles on the local electric field is accounted for by including polarisability in the model. An example of a polarisable model is the shell model, where a charge is attached to the atom by a spring, hence the dipole of the atom reacts to changes in the local electric field.

4.5 Neural Network Potentials

Behler and Parrinello [33] presented a new scheme for generating interatomic potentials using neural networks that are trained to reproduce quantum mechanical data. The main assumption of the model is that the total energy of an atomic system can be described as a sum of atomic contributions

$$E = \sum_i E_i, \tag{4.57}$$

where each individual term E_i depends only on the configuration of the neighbouring atoms within a given cutoff distance. This local environment is represented using a set of symmetry functions

$$G_i^{1\alpha} = \sum_{j \neq i} \exp[-\eta_\alpha (r_{ij} - r_{s\alpha})^2] f_{\mathrm{cut}}(r_{ij}) \tag{4.58}$$

$$\begin{aligned} G_i^{2\beta} = 2^{1-\zeta_\beta} \sum_{j \neq i} \sum_{k \neq i} (1 + \lambda_\beta \cos \theta_{ijk})^{\zeta_\beta} \\ \exp\left[-\eta_\beta \left(r_{ij}^2 + r_{ik}^2 + r_{jk}^2\right)\right] f_{\mathrm{cut}}(r_{ij}) f_{\mathrm{cut}}(r_{ik}) f_{\mathrm{cut}}(r_{jk}), \end{aligned} \tag{4.59}$$

where the cutoff function is

$$f_{\mathrm{cut}}(r) = \begin{cases} \frac{1}{2} + \frac{1}{2}\cos\left(\frac{\pi r}{r_{\mathrm{cut}}}\right) & \text{if} \quad r_{ij} \leq r_{\mathrm{cut}} \\ 0 & \text{if} \quad r_{ij} > r_{\mathrm{cut}} \end{cases}. \tag{4.60}$$

Thus the atomic local energies E_i depend on the set of symmetry variables $\{G_i^{1\alpha}, G_i^{2\beta}\}$ in an unknown way. Instead of trying to find a parametric model for this function, Behler and Parrinello used non-parametric regression via neural networks. The input data used to perform the regression is a set of total energies from reference calculations, in this case these were Density Functional Theory calculations of different configurations of bulk silicon. The parameters in the layers of the neural network were optimised such that the difference between the reference energies and the energies predicted by the neural network is minimal. The resulting potential can then be used to describe an arbitrary number of silicon atoms. For each atom, the symmetry variables are first determined, then these are fed to the neural network and the neural network predicts the atomic energies, which are added together to obtain the total energy.

4.6 Gaussian Approximation Potentials

Our aim is to formulate a generic interatomic potential, which can be reliably used in a wide variety of applications. Arguably, Quantum Mechanics is such an interatomic potential, as it provides ab initio data that, to our current knowledge, is ultimately correct to the extent that any inaccuracies are due to the limitation of the Born–Oppenheimer approximation or the employed quantum mechanical model. The great advantage of quantum mechanical methods is that they have true and proven predictive power, whereas classical potentials can be regarded as parametric regression formulas that, in general, cannot be used outside their fitting regime, which usually cannot be unambiguously classified. However, the solution of quantum mechanical equations is computationally expensive, which limits the use of Quantum Mechanics to a modest number of atoms and a few nanoseconds of simulation time—woefully inadequate for biomolecular and nanotechnological applications.

As in the case of other interatomic potentials, we base Gaussian Approximation Potentials on the assumption that the total energy of the system can be written as a sum of two terms: the first is a local, atomic contribution and the second is the long-range, electrostatic part

$$E = \sum_{i}^{\text{atoms}} \varepsilon_i + \frac{1}{2} \sum_{i<j}^{\text{atoms}} \hat{L}_i \hat{L}_j \frac{1}{r_{ij}}, \tag{4.61}$$

where the operator $\hat{L}$ can be written as

$$\hat{L}_i = q_i + \mathbf{p}_i \cdot \nabla_i + \mathbf{Q}_i : \nabla_i \nabla_i + \cdots, \tag{4.62}$$

and q_i, $\mathbf{p}_i$ and $\mathbf{Q}_i$ denote the charge, dipole and quadrupole of the ith atom, respectively. We formulate the locality of the atomic energy contributions as

$$\varepsilon_i \equiv \varepsilon(\{\mathbf{r}_{ij}\}), \tag{4.63}$$

where only the relative positions $\mathbf{r}_{ij}$ of the neighbouring j atoms within a spherical cutoff are considered. In atomic systems, for which charge transfer between atoms and polarisation effects are negligible, we can simply drop the second term in Eq. 4.61. We note that short-range, well screened electrostatic effects can be implicitly merged into the first term in Eq. 4.61 without great sacrifices in accuracy.

The strict localization of ε enables the independent computation of atomic energies.

The central challenge in the development of interatomic potentials is finding the form of $\varepsilon(\{\mathbf{r}_{ij}\})$. In our approach, we do not make any prior assumptions about the functional form of the potential. Instead, we use non-parametric, non-linear regression in the form of a Gaussian Process to find the function values at arbitrary values. In the regression, quantum mechanical data, such as total energies and atomic forces are used as evidence. Gaussian Approximation Potentials can be regarded as interpolation of the quantum mechanical potential energy surface. Moreover, the Gaussian Process framework allows us to to build into the model a strong bias, namely, that the atomic energy function is smooth.

The advantage of Gaussian Approximation Potentials is that they are very flexible. In contrast to analytic potentials, the accuracy of Gaussian Approximation Potentials can be improved by adding more quantum mechanical data at various points in configurational space without changing the fit globally. As the Gaussian Process predicts its own accuracy, it is possible to use it as a "learn on the fly" method, i.e. if the predicted variance of the energy of the force in the case of a new configuration is higher than a pre-set tolerance, the energy and forces for the new configuration can be calculated using Quantum Mechanics, then the obtained data is added to the database in order to improve the fit. The flexibility of the fit ensures that the best possible fit is achieved for any given data.

The Gaussian Approximation Potential scheme is similar to the Neural Network potentials introduced by Behler and Parinello [33], as both uses non-linear, non-parametric regression instead of fixed analytic forms. However, the representation of the atomic environments in GAP is complete and the Gaussian Process uses energies and forces for regression. Moreover, the training of the neural network involves the optimisation of the weights, whereas the training in the case of Gaussian Process is a simple matrix inversion.

4.6.1 Technical Details

The atomic energy function ε depends on the atomic neighbourhood, but it is invariant under rotation, translation and permutation of the atoms. One of the key ideas in the present work is to represent atomic neighbourhoods in a transformed system of coordinates that accounts for these symmetries. Ideally, this mapping should be one-to-one: mapping different neighbourhood configurations to the same coordinates would introduce systematic errors into the model that cannot be improved by adding more quantum mechanical data. In Sect. 2.3 in Chap. 2 we described a number of transformations that can be adapted to construct an invariant neighbourhood representation. For our work, we have chosen the four dimensional bispectrum elements. In order to ensure that the representation is continuous in space, we modified the atomic density in Eq. 2.18 to

$$\rho_i(\mathbf{r}) = \delta(\mathbf{r}) + \sum_j \delta(\mathbf{r} - \mathbf{r}_{ij}) f_{\mathrm{cut}}(r_{ij}), \tag{4.64}$$

where f_{cut} is a cutoff function, in our case

$$f_{\mathrm{cut}}(r) = \begin{cases} 0 & \text{if} \quad r > r_{\mathrm{cut}} \\ 1/2 + \cos(\pi r / r_{\mathrm{cut}})/2 & \text{if} \quad r \leq r_{\mathrm{cut}} \end{cases}. \tag{4.65}$$

In Quantum Mechanics, atomic energies are not directly accessible, only the total energy of a configuration and the forces on each atom can be determined. The forces contain cross-terms of the derivatives of the local energies. The force on atom i can be obtained by differentiating the total energy with respect to the Cartesian coordinates of atom i, written as

$$f_{i\alpha} = \frac{\partial E}{\partial r_{i\alpha}} = \sum_j^{\mathrm{atoms}} \frac{\partial \varepsilon_j}{\partial r_{i\alpha}}. \tag{4.66}$$

As $\varepsilon \equiv 0$ for any $r_{ij} > r_{\{\mathrm{cut}}$, this summation only runs over the N_i neighbours of atom i. The atomic energies depend directly on the bispectrum elements, which are determined by the neighbourhood, thus the force becomes

$$f_{i\alpha} = \sum_j^{N_i} \sum_k \left(\frac{\partial \varepsilon}{\partial b_k} \right)_{\mathbf{b}_j} \frac{\partial b_k}{\partial r_{i\alpha}}, \tag{4.67}$$

where b_k is the kth element of the bispectrum vector, and $\mathbf{b}_j$ is the bispectrum of atom j. Therefore we can substitute total energy observations in the form of sums of atomic energies, and forces, in the form of sums of derivatives of atomic energies, directly in the formulae shown in Sects. 3.2.3 and 3.2.4 in Chap. 3.

If N is the number of teaching points, the computational resources required for Gaussian Process regression scales as N^3 for training and as N for predicting values and as N^2 for predicting variances. Due to the fact that we cannot add single atomic energy observations to the database, only total energies or forces, the size of the training set and therefore the computational costs would grow enormously. For example, if we intend to add configurations with defects to a database that up to this point contains data for bulk atoms only, we have to add all the atomic neighbourhoods in the configuration that contains the defect, despite of the fact that most of them are redundant because they incorporate the bulk data that is already in the database. Similarly, a single configuration can contain many correlated neighbourhoods.

A possible solution for this problem was given by Snelson and Ghahramani [34] and it was described in Sect. 3.2.5 in Chap. 3. By choosing M sparse points from the complete training set, the computational resources required for the training process scale as NM^2, while the cost of the prediction of function values and variances scales as M and M^2, respectively.

4.6.2 Multispecies Potentials

It is possible to extend the scope of Gaussian Approximation Potentials to cases where there are more than one atomic species present in the system. There are two main differences with respect to the method described above for monoatomic potentials. On the one hand, the different species have to be distinguished in the atomic neighbourhood while retaining the rotational and permutational invariance, and, on the other hand, charge transfer between different types of atoms might occur, in which case the long-range interactions have to be taken in account. The latter is not necessary in every multispecies system, for example, in hydrocarbons or metallic alloys there are no significant long-range interactions present [35].

By modifying the atomic density function in Eq. 2.18 as in Eq. 2.68:

$$\rho_i(\mathbf{r}) = s_i\delta(\mathbf{0}) + \sum_j s_j\delta(\mathbf{r} - \mathbf{r}_{ij}), \tag{4.68}$$

where the different species are distinguished by the different weights of the Dirac-delta functions. The bispectrum of ρ_i remains invariant to the global rotation of the atomic neighbourhood and to permutations of atoms of the same species.

In this study, we have not developed any potentials that contain electrostatics explicitly, but there is good evidence [36], that electrostatic parameters, such as charges and multipoles can be obtained from electronic structure calculations. It is possible to fix these parameters, but in general, the charges and multipoles will be determined by the local neighbourhood and the local electric field, and so these effects must be incorporated any accurate potential. This branch of our research awaits implementation.

References

1. D.W. Brenner, Phys. Stat. Sol. (b) **217**, 23 (2000)
2. M. Finnis, *Interatomic Forces in Condensed Matter*. (Oxford University Press, Oxford, 2003)
3. A. Szabo, N. S. Ostlund, *Modern Quantum Chemistry: Introduction to Advanced Electronic Structure Theory*. (Dover Publications, New York, 1996)
4. W. Kohn, L.J. Sham, Phys. Rev. **140**, 1133 (1965)
5. S.J. Clark et al., Zeit. Krist. **220**, 567 (2005)
6. N.B. Wilding, A.D. Bruce, Phys. Rev. Lett. **85**, 5138 (2000)
7. W.G. Hoover, F.H. Ree, J. Chem. Phys. **49**, 3609 (1968)
8. A. Fortini, M. Dijkstra, J. Phys. Cond. Mat. **18**, L371 (2008)
9. P.N. Pusey, W. van Megen, Nature **320**, 340 (1986)
10. J.E. Jones, Proc. R. Soc. Lond. Ser. A **106**, 463 (1924)
11. M.P. Allen, D.J. Tildesley, *Computer Simulation of Liquids*. (Oxford University Press, Oxford, 1987)
12. W.L. Jorgensen, *Encyclopedia of Computational Chemistry*. (Wiley, New York, 1998)
13. K.P. Jensen, W.L. Jorgensen, J. Chem. Theo. Comp. **2**, 1499 (2006)
14. H.J.C. Berendsen, J.R. Grigera, T.P. Straatsma, J. Phys. Chem. **91**, 6269 (1987)
15. W.L. Jorgensen, J. Chandrasekhar, J.D. Madura, R.W. Impey, M.L. Klein, J. Chem. Phys. **79**, 926 (1983)
16. Y.-X. Yu, G.H. Gao, Int. J. ThermoPhys. **21**, 57 (2000)
17. L.B. Páartay, A.P. Bartóok, G. Csáanyi, submitted (2009)
18. J. Hernandez-Rojas, D.J. Wales, J. Non-Cryst. Solids **336**, 218 (2004)
19. J.R. Morris, X. Song, J. Chem. Phys. **116**, 9352 (2002)
20. M.S. Daw, M.I. Baskes, Phys. Rev. Lett. **50**, 1285 (1983)
21. F. Ercolessi, J.B. Adams, Europhys. Lett. **26**, 583 (1994)
22. M.I. Baskes, Phys. Rev. Lett. **59**, 2666 (1987)
23. T.J. Lenosky et al., Mod. Sim. Mat. Sci. Eng. **8**, 825 (2000)
24. C.D. Taylor, Phys. Rev. B **80**, 024104 (2009)
25. J. Tersoff, Phys. Rev. B **38**, 9902 (1988)
26. P. Alinaghian, P. Gumbsch, A.J. Skinner, D.G. Pettifor, J. Phys. Cond. Mat. **5**, 5795 (1993)
26. P.G. Cummins, D.A. Dunmur, R.W. Munn, N.R.J, Acta Crystallogr. Sect. A **32**, 847 (1976)
28. H.W. Horn et al., J. Chem. Phys. **120**, 9665 (2004)
29. Q. Chen, L. Cai, S. Duan, D. Chen, J. Phys. Chem. Sol. **65**, 1077 (2004)
30. W.J. Mortier, S.K. Ghosh, S. Shankar, J. Am. Chem. Soc. **108**, 4315 (2002)
31. A.K. Rappe, W.A. Goddard, J. Phys. Chem. **95**, 3358 (2002)
32. C. Sagui, L.G. Pedersen, T.A. Darden, J. Chem. Phys. **120**, 73 (2004)
33. J. Behler, M. Parrinello, Phys. Rev. Lett. **98**, 146401 (2007)
34. E. Snelson, Z. Ghahramani, Sparse gaussian processes using pseudo-inputs, in *Advances in Neural Information Processing Systems 18*, ed. by Y. Weiss, B. Schölkopf, J. Platt (MIT Press, Cambridge, 2006), pp. 1257–1264
35. D.W. Brenner, Phys. Rev. B **42**, 9458 (1990)
36. M. Wilson, S. Jahn, P.A. Madden, J. Phys.: Cond. Mat. **16**, (2004)

Chapter 5
Computational Methods

5.1 Lattice Dynamics

5.1.1 Phonon Dispersion

Crystalline materials are composed of periodic replicas of unit cells. In our case, the unit cell is a parallelepiped defined by the edge vectors $\mathbf{a}_1$, $\mathbf{a}_2$ and $\mathbf{a}_3$. The volume of the unit cell is the absolute value of determinant of the lattice matrix $\mathbf{A} = [\mathbf{a}_1, \mathbf{a}_2, \mathbf{a}_3]$, which is nonzero, as the column vectors of the matrix are linearly independent. The smallest unit cell is called the primitive cell. The positions $\mathbf{r}_j^0$ of the atoms in the primitive cell form the basis of the crystal.

The crystal is built by translating the primitive cell by all the translation vectors

$$\mathbf{R}_\mathbf{l} = l_1\mathbf{a}_1 + l_2\mathbf{a}_2 + l_3\mathbf{a}_3, \tag{5.1}$$

where l_1, l_2 and l_3 are integers. Hence the equilibrium position of the ith atom in the crystal can be written as

$$\mathbf{r}_i^0 = \mathbf{r}_{\mathbf{l}j}^0 = \mathbf{r}_j^0 + \mathbf{R}_\mathbf{l}. \tag{5.2}$$

At finite temperature, atoms vibrate around their equilibrium positions, and their displacement can be described by a small vector $\mathbf{u}$. The actual position of an atom is given by

$$\mathbf{r}_{\mathbf{l}j} = \mathbf{r}_{\mathbf{l}j}^0 + \mathbf{u}_{\mathbf{l}j}. \tag{5.3}$$

The total potential energy ϕ of the crystal is a function of the positions of the atoms. The Taylor-expansion of the potential energy is

$$\phi = \phi_0 + \sum_{\mathbf{l},j,\alpha} \phi_{\mathbf{l}j\alpha} u_{\mathbf{l}j\alpha} + \frac{1}{2}\sum_{\mathbf{l},j,\alpha}\sum_{\mathbf{l}',j',\alpha'} \phi_{\mathbf{l}j\alpha,\mathbf{l}'j'\alpha'} u_{\mathbf{l}j\alpha} u_{\mathbf{l}'j'\alpha'} + \cdots, \tag{5.4}$$

A. Bartók-Pártay, *The Gaussian Approximation Potential*, Springer Theses,
DOI: 10.1007/978-3-642-14067-9_5, © Springer-Verlag Berlin Heidelberg 2010

where ϕ_0 is the equilibrium energy. The first term in Eq. 5.4 is the related to the force through

$$\phi_{\mathbf{l}j\alpha} = \frac{\partial \phi}{\partial u_{\mathbf{l}j\alpha}} = -f_{\mathbf{l}j\alpha}. \tag{5.5}$$

This term is zero, because we perform the Taylor expansion around the minimum. The second term contains the harmonic force constants, given by

$$\phi_{\mathbf{l}j\alpha,\mathbf{l}'j'\alpha'} = \frac{\partial^2 \phi}{\partial u_{\mathbf{l}j\alpha} \partial u_{\mathbf{l}j\alpha}}. \tag{5.6}$$

In the harmonic approximation, higher order terms in the Taylor-expansion are neglected. Newton's equations of motion are therefore written as

$$m_j \ddot{u}_{\mathbf{l}j\alpha} = \sum_{\mathbf{l}'j'\alpha'} \phi_{\mathbf{l}j\alpha,\mathbf{l}'j'\alpha'} u_{\mathbf{l}'j'\alpha'}, \tag{5.7}$$

which have wavelike solutions

$$\mathbf{u}_{\mathbf{l}j}(t) = \frac{1}{\sqrt{Nm_j}} \sum_{\mathbf{k}v} A(\mathbf{k}, v)\mathbf{e}(\mathbf{k}, v, j) \exp\left[i\left(\mathbf{k}\mathbf{r}^0_{\mathbf{l}j} - \omega(\mathbf{k}, v)t\right)\right]. \tag{5.8}$$

Substituting 5.8 in 5.7, we obtain the eigenvalue equation

$$\omega^2(\mathbf{k}) e_\alpha(\mathbf{k}i, v, j) = \sum_{\alpha' j' v'} D_{\alpha\alpha'}\begin{pmatrix} \mathbf{k} \\ jv, j'v' \end{pmatrix} e_{\alpha'}(\mathbf{k}, v', j'), \tag{5.9}$$

where $\mathbf{D}$ is the dynamical matrix, the Fourier transform of the force constant matrix:

$$D_{\alpha\alpha'}\begin{pmatrix} \mathbf{k} \\ jv, j'v' \end{pmatrix} = \frac{1}{m_j m_{j'}} \sum \phi_{\mathbf{l}j\alpha,\mathbf{l}'j'\alpha'} \exp\left[i\mathbf{k}\left(\mathbf{r}^0_{\mathbf{l}j} - \mathbf{r}^0_{\mathbf{l}'j'}\right)\right]. \tag{5.10}$$

Non-trivial solutions of Eq. 5.9 can be found by solving the secular determinant

$$|\mathbf{D}(\mathbf{k}) - \omega^2 \mathbf{I}| = 0, \tag{5.11}$$

where the solutions are the frequencies of different phonon modes at wavevector $\mathbf{k}$. Substituting these solutions into 5.9, the mode eigenvectors can also be obtained, and these correspond to the normal modes of the vibrations. A more complete discussion of lattice dynamics can be found, for example, in [1].

In our work, we first constructed a large supercell from the primitive cell, then perturbed each atom in the original $\mathbf{l} = (0, 0, 0)$ cell by a small amount along the coordinate axes and calculated the forces on the atoms in the perturbed supercell. We obtained an approximate force constant matrix by the numerical differentiation of the forces, which we Fourier-transform to obtain the dynamical matrix. This procedure can be performed using any interatomic potential model, although using

Quantum Mechanics can be particularly expensive in the case of large supercells, i.e. for small wavenumbers. However, this large computational cost in DFT can be avoided by calculating phonon dispersion relations using Density Functional Perturbation Theory, as described in [2].

5.1.2 Molecular Dynamics

Alternatively, the phonon frequencies can also be obtained from molecular dynamics runs [3]. The relative displacements $\mathbf{u}_{\mathrm{l}j}$ in Eq. 5.8 can be Fourier-transformed, leading to

$$\epsilon_{\mathbf{k}j}(t) = \frac{1}{N_{\mathrm{cell}}} \sum_j \sum_{\mathrm{l}} \exp(-i\,\mathbf{k}\,\mathbf{R}_{\mathrm{l}})\mathbf{u}_{\mathrm{l}j} \propto \sum_v \exp(-i\omega(\mathbf{k}, v)t), \tag{5.12}$$

where N_{cell} is the number of primitive cells in the supercell. Fourier-transforming equation 5.12 to frequency space gives

$$\epsilon_{\mathbf{k}}(\omega) \propto \sum_v \delta(\omega - \omega_{\mathbf{k},v}). \tag{5.13}$$

The spectral analysis of $\epsilon_{\mathbf{k}}(\omega)$, i.e. finding sharp peaks in the power spectrum

$$P_{\mathbf{k}j} \equiv |\epsilon_{\mathbf{k}j}(\omega)|^2 \tag{5.14}$$

gives the phonon frequencies.

The advantage of this method is that it can be used for more complicated systems, where explicit calculation of the full dynamical matrix would be extremely expensive. Furthermore, we can calculate the temperature dependence of the phonon spectrum by simply performing molecular dynamics simulations at different temperatures. The temperature dependence of the phonon spectrum is due to anharmonic effects, i.e., at larger displacements when terms higher than second order contribute to the potential energy in Eq. 5.4.

5.1.3 Thermodynamics

The quantum mechanical solution of a system of harmonic oscillators [1] states that the allowed energies of a phonon mode labelled by $\mathbf{k}$ and v are

$$E_{\mathbf{k}v} = \left(\frac{1}{2} + n\right)\omega(\mathbf{k}, v), \tag{5.15}$$

where $\hbar$ is the reduced Planck constant, and n is a non-negative integer. The canonical partition function of a system can be calculated as

$$Z = \sum_j \exp(-\beta E_j), \tag{5.16}$$

where E_j is the energy of the jth state and $\beta = \frac{1}{k_B T}$. Substituting 5.15 into this expression, we obtain

$$Z_{\mathrm{vib.}} = \prod_{\mathbf{k},\nu} \left[\sum_{n_{\mathbf{k}\nu}=0}^{\infty} \exp\left(-\beta \left(\frac{1}{2} + n_{\mathbf{k}\nu} \right) \hbar\omega(\mathbf{k},\nu) \right) \right] \tag{5.17}$$

which can be simplified by using

$$\sum_{n=0}^{\infty} \exp(-nx) = \frac{1}{1-\exp(-x)} \tag{5.18}$$

to

$$Z_{\mathrm{vib.}} = \prod_{\mathbf{k},\nu} \frac{\exp(-\beta\hbar\omega(\mathbf{k},\nu)/2)}{1-\exp(-\beta\hbar\omega(\mathbf{k},\nu))}. \tag{5.19}$$

In the case of a crystal, the total partition function is

$$Z = \exp(-\beta\phi_0) Z_{\mathrm{vib.}}. \tag{5.20}$$

The partition function can be used to obtain all thermodynamic quantities. For example, the free-energy can be obtained as

$$F = -k_B T \ln Z \tag{5.21}$$

$$= \phi_0 + k_B T \sum_{\mathbf{k},\nu} \ln[2\sinh(\beta\hbar\omega(\mathbf{k},\nu)/2)], \tag{5.22}$$

and the internal energy is

$$U = \frac{1}{Z}\frac{\partial Z}{\partial \beta} \tag{5.23}$$

$$= \phi_0 + \sum_{\mathbf{k},\nu} \hbar\omega(\mathbf{k},\nu) \left(\frac{1}{2} + \frac{1}{\exp(-\beta\hbar\omega(\mathbf{k},\nu)) - 1} \right). \tag{5.24}$$

This result leads us to a rather crude method for approximating the real temperature in the case of a classical molecular dynamics run [4]. We equate the kinetic energy $E_{\mathrm{kin.}}(T_{\mathrm{MD}})$ to the quantum mechanical vibration energy $U_{\mathrm{vib.}}(T_{\mathrm{QM}})$ and find the temperature T_{QM} when $U_{\mathrm{vib.}}(T_{\mathrm{QM}}) = E_{\mathrm{kin.}}(T_{\mathrm{MD}})$. In the high temperature limit $T_{\mathrm{QM}} = T_{\mathrm{MD}}$, but this expression allows us to relate results from low-temperature molecular dynamics runs to experimental values.

The constant-volume heat capacity is defined as

$$c_V = \frac{\partial U}{\partial T}, \tag{5.25}$$

which, in the case of harmonic crystals, can be calculated as

$$C_V = \sum_{\mathbf{k},\nu} c_{\mathbf{k},\nu}, \tag{5.26}$$

where $c_{\mathbf{k},\nu}$ is the contribution to the specific heat from mode $(\mathbf{k},\nu)$

$$c_{\mathbf{k},\nu} = k_B \left(\frac{\hbar\omega(\mathbf{k},\nu)}{k_B T}\right)^2 \frac{\exp(\beta\hbar\omega(\mathbf{k},\nu))}{[\exp(\beta\hbar\omega(\mathbf{k},\nu)) - 1]^2}. \tag{5.27}$$

The volumetric thermal expansion coefficient can also be calculated from the free energy. The thermal expansion coefficient is defined as

$$\alpha = \frac{1}{V}\left(\frac{\partial V}{\partial T}\right)_p = \frac{1}{V}\left(\frac{\partial V}{\partial p}\right)_T \left(\frac{\partial p}{\partial T}\right)_V = \kappa_T \left(\frac{\partial p}{\partial T}\right)_V, \tag{5.28}$$

where κ_T is the isothermal compressibility. The pressure is given by

$$p = -\left(\frac{\partial F}{\partial V}\right)_T, \tag{5.29}$$

which leads to the expression

$$\alpha = \frac{\kappa_T}{V} \sum_{\mathbf{k},\nu} \gamma_{\mathbf{k},\nu} c_{\mathbf{k},\nu}, \tag{5.30}$$

where $\gamma_{\mathbf{k},\nu}$ are the k-vector dependent Grüneisen parameters

$$\gamma_{\mathbf{k},\nu} = -\frac{V}{\omega(\mathbf{k},\nu)} \partial\omega(\mathbf{k},\nu)\partial V = \frac{\partial \ln \omega(\mathbf{k},\nu)}{\partial \ln V}, \tag{5.31}$$

which describe the dependence of the phonon frequencies on the lattice volume. The linear thermal expansion can be obtained in a similar way and the derivation can be easily extended to non-isotropic cases.

We note that through the Grüneisen parameters anharmonic corrections of the potential energy are involved in the thermal expansion coefficient. The approximation that the vibrational free-energy function depends on the volume of the crystal through the change of the phonon frequencies described by the first-order approximation

$$\omega(\mathbf{k},\nu,V) = \omega(\mathbf{k},\nu,V_0) + \frac{\partial\omega(\mathbf{k},\nu)}{\partial V}\Delta V \tag{5.32}$$

is usually referred to as the quasi-harmonic approximation [1].

At low temperatures, if most of the anharmonic effects are due to lattice expansion, the quasi-harmonic approximation can be successfully applied. However, if the average displacement of the atoms is so large that the potential energy cannot be approximated by quadratic terms anymore, the approximation fails. In such cases, we can use a classical simulation method such as molecular dynamics to sample the phase space and calculate observables using these samples. We should note that this is strictly valid only in case of high temperatures, where $T_{\mathrm{MD}} \approx T_{\mathrm{QM}}$.

However, if the anharmonic effects are large even at low temperatures, precise results can be obtained by methods that treat the quantum character of the nuclei explicitly, for example by path-integrals [5] or explicitly solving the nuclear Schrödinger equation [6]. Path-integral methods have been successfully used to calculate the partition function of semiconductor crystals [7] and hydrogen impurity in metals [6]. Explicit solution of the nuclear Schrödinger equation is routinely performed in the case of molecules [8] by using the system of eigenfunctions of the harmonic solution to expand the wavefunction.

References

1. M.T. Dove, *Introduction to Lattice Dynamics*. (Cambridge University Press, Cambridge, 1993)
2. S. Baroni, S. de Gironcoli, A. Dal Corso, P. Giannozzi, Rev. Mod. Phys. **73**, 515 (2001)
3. T.A. Arias, M.C. Payne, J.D. Joannopoulos, Phys. Rev. B **45**, 1538 (1992)
4. C.Z. Wang, C.T. Chan, K.M. Ho, Phys. Rev. B **42**, 11276 (1990)
5. L.D. Fosdick, H.F. Jordan, Phys. Rev. **143**, 58 (1966)
6. M.J. Gillan, Phil. Mag. A **58**, 257 (1988)
7. C.P. Herrero, Phys. Rev. B **63**, 024103 (2000)
8. W.D. Allen et al., Chem. Phys. **145**, 427 (1990)

Chapter 6
Results

6.1 Atomic Energies

The total energy in Quantum Mechanics is a global property of the system consisting of N atoms and depends on $3N$–6 variables, namely, the coordinates of the atoms. However, all interatomic potentials are based on the assumption that the energy can be written as a sum of atomic or bond energies, which are local and if appropriate, a long-range electrostatic component. In our work, we intend to estimate the atomic energies by a regression scheme based directly on quantum mechanical data. If there were a way to extract atomic energies directly from quantum mechanical calculations, these could be used in the regression. Firstly, we consider ideas that lead to such atomic energies.

In fact, the existence of atomic energies can be justified by showing that the force acting on an atom does not change *significantly* if the position of another atom that is *far enough away* is perturbed. This statement can be formulated as

$$\nabla^n_{\mathbf{x}_j} \nabla_{\mathbf{x}_i} E \to 0 \quad \text{as } |\mathbf{x}_j - \mathbf{x}_i| \to \infty, \quad \text{for } \forall n, \tag{6.1}$$

which we refer to as the "strong locality assumption".

6.1.1 Atomic Expectation Value of a General Operator

The basic idea in the derivation of atomic properties in Quantum Mechanics is partitioning the total expectation value of an arbitrary operator by using a suitable atomic basis set. This is a generalisation of the Mulliken charge partitioning scheme. We consider a system of non interacting electrons moving in an effective potential V_{eff}, which is the case in DFT. Thus the expectation value of a general operator $\hat{O}$ is

A. Bartók-Pártay, *The Gaussian Approximation Potential*, Springer Theses,
DOI: 10.1007/978-3-642-14067-9_6, © Springer-Verlag Berlin Heidelberg 2010

$$\langle O\rangle = \sum_i f_i \langle \psi_i|\hat{O}|\psi_i\rangle, \tag{6.2}$$

where f_i is the occupation number of the single-electron orbital ψ_i. If ψ_i is expressed in an atomic basis $\{\phi_\alpha\}$ in the form

$$\psi_i = \sum_\alpha N_i^\alpha \phi_\alpha, \tag{6.3}$$

we can write Eq. 6.2 as

$$\langle O\rangle = \sum_i \sum_{\alpha\beta} f_i N_i^\alpha \left(N_i^\beta\right)^* \langle \phi_\beta|\hat{O}|\phi_\beta\rangle. \tag{6.4}$$

Introducing the density kernel $\mathbf{K}$ as

$$K^{\alpha\beta} = \sum_i f_i N_i^\alpha \left(N_i^\beta\right)^* \tag{6.5}$$

and the matrix of operator $\hat{O}$ as

$$O_{\beta\alpha} = \langle \phi_\beta|\hat{O}|\phi_\alpha\rangle \tag{6.6}$$

we obtain

$$\langle O\rangle = \sum_{\alpha\beta} K^{\alpha\beta} O_{\beta\alpha} = \sum_\alpha (\mathbf{KO})_{\alpha\alpha} = \mathrm{Tr}(\mathbf{KO}). \tag{6.7}$$

Each basis function ϕ_α belongs to a certain atom, thus we use the partitioning

$$\langle O\rangle_A = \sum_{\alpha\in A} (KO)_{\alpha\alpha}, \tag{6.8}$$

which conserves the total value

$$\langle O\rangle = \sum_A \langle O\rangle_A. \tag{6.9}$$

6.1.1.1 Mulliken Charges

The total number of electrons is obtained by setting the operator to $\hat{O} = 1$:

$$N = \sum_i f_i \langle \psi_i|\psi_i\rangle = \sum_i f_i = \sum_{\alpha,\beta} K^{\alpha\beta} \int \phi_\alpha(\mathbf{r})\phi_\beta^*(\mathbf{r}) \tag{6.10}$$

which leads to the well-known expression for the Mulliken-charges

$$N_A = \sum_{\alpha \in A} (\mathbf{KS})_{\alpha\alpha}, \tag{6.11}$$

where the elements of the overlap matrix are defined by

$$S_{\alpha\beta} = \int \phi_\alpha^*(\mathbf{r}) \phi_\beta(\mathbf{r}) \tag{6.12}$$

6.1.2 Atomic Energies

Substituting the Hamiltonian operator $\hat{H}$ into Eq. 6.8, we obtain a possible definition for the atomic energies. In the case of Density Functional Theory, the operators can be formulated as follows.

The total energy can be written as

$$E[\rho] = T_s + E_{\mathrm{H}}[\rho] + E_{\mathrm{xc}}[\rho] + E_{Ze}[\rho] + E_{ZZ}. \tag{6.13}$$

The independent-particle kinetic energy T_s is given by

$$T_s = -\frac{1}{2} \sum_i^{\mathrm{elec.}} f_i \langle \psi_i | \Delta_i | \psi_i \rangle, \tag{6.14}$$

thus we need to substitute $\hat{O} = -\frac{1}{2} \sum_i \Delta_i$ and the matrix elements $\sum_i \langle \phi_\alpha | \Delta_i | \phi_\beta \rangle$ have to be calculated to obtain the atomic kinetic energy.

The Hartree energy is defined by the Eq.

$$E_{\mathrm{H}}[\rho] = \frac{1}{2} \int \int \mathbf{dr}\,\mathbf{dr}' \frac{\rho(\mathbf{r})\,\rho(\mathbf{r}')}{|\mathbf{r} - \mathbf{r}'|}, \tag{6.15}$$

which we rewrite as

$$E_{\mathrm{H}}[\rho] = \frac{1}{2} \int \mathbf{dr}\,\rho(\mathbf{r}) V_{\mathrm{H}}(\mathbf{r}) = \frac{1}{2} \sum_i f_i \langle \psi_i | \hat{V}_H | \psi_i \rangle, \tag{6.16}$$

where the Hartree-operator can be obtained as

$$V_{\mathrm{H}}(\mathbf{r}) = \frac{1}{2} \int \mathbf{dr}' \frac{\rho(\mathbf{r}')}{|\mathbf{r} - \mathbf{r}'|}. \tag{6.17}$$

Similarly, the interaction between electrons and nuclei is given by

$$E_{Ze}[\rho] = \int \mathbf{dr}\,\rho(\mathbf{r}) V_{\mathrm{ext}}, \tag{6.18}$$

and the exchange-correlation energy is

$$E_{xc}[\rho] = \int \mathrm{d}\mathbf{r}\, \rho(\mathbf{r})\epsilon_{xc}[\rho(\mathbf{r})]. \tag{6.19}$$

Hence the operators V_{ext} and $\epsilon_{xc}[\rho(\mathbf{r})]$ are required to calculate the matrix elements of the external energy matrix and the exchange-correlation energy matrix.

6.1.3 Atomic Multipoles

In general, the multipole coefficients of an arbitrary charge distribution $\rho(\mathbf{r})$ can be obtained as

$$\mu_{\xi_1,\xi_2,\ldots,\xi_l} = \frac{1}{l!}\int \mathrm{d}\mathbf{r}\, x_{\xi_1}x_{\xi_2},\ldots,x_{\xi_l}\,\rho(\mathbf{r}), \tag{6.20}$$

where $x_{\xi_1}, x_{\xi_2}, \ldots, x_{\xi_l}$ are the Cartesian coordinates. This definition can be regarded as an expectation value of the general position operator $\hat{X}$, therefore it can be substituted into Eq. 6.8, to produce the definition of atomic multipoles:

$$\mu_{A,\xi_1,\xi_2,\ldots,\xi_l} = \frac{1}{l!}\sum_{\alpha\in A}\sum_{\beta} K^{\alpha\beta}\int \mathrm{d}\mathbf{r}\, x_{\xi_1}x_{\xi_2},\ldots,x_{\xi_l}\,\phi_\alpha(\mathbf{r})\phi^*_\beta(\mathbf{r}), \tag{6.21}$$

where x_{ξ_l} is measured from atom A.

It is interesting to note that the expression for atomic multipoles in Eq. 6.21 can be obtained by defining the atomic charge density ρ_A as

$$\rho_A(\mathbf{r}) = \sum_{\alpha\in A} K^{\alpha\beta}\phi_\alpha(\mathbf{r})\phi^*_\beta(\mathbf{r}). \tag{6.22}$$

This definition of the atomic charge density is consistent with general physical considerations, for example it gives the total electron density when summed for all atoms:

$$\rho(\mathbf{r}) = \sum_{A}\rho_A(\mathbf{r}). \tag{6.23}$$

6.1.4 Atomic Energies from ONETEP

ONETEP [1], the order-N electronic total energy package is a numerical implementation of Density Functional Theory. Unlike usual implementations of DFT, the computational resources required for the calculation of the energy of a particular atomic system scales linearly with the number of electrons, which makes it exceptionally efficient in investigations of large systems. However, in our work we exploited another feature of ONETEP, namely, that it uses local basis functions.

6.1.4.1 Wannier Functions

The electronic structure of periodic crystalline solids is usually represented by Bloch orbitals $\psi_{n\mathbf{k}}$, where n and $\mathbf{k}$ are quantum numbers of the band and crystal momentum, respectively. The Bloch states are eigenfunctions of the Hamiltonian of the crystal, obeying the same periodicity. Because of the fact that they are usually highly delocalised, it is often difficult to deduce local properties from Bloch orbitals, for instance, bonding between atoms or atomic charges.

An equivalent representation of the electronic structure is provided by Wannier functions [2], which are connected to the Bloch orbitals via a unitary transformation. Denoting the Wannier functions of band n of cell $\mathbf{R}$ by $w_n(\mathbf{r}-\mathbf{R})$, we express the transformation as follows:

$$w_n(\mathbf{r}-\mathbf{R}) = \frac{V}{8\pi^3}\int \mathrm{d}\mathbf{k}\, e^{-i\mathbf{k}\mathbf{R}}\, \psi_{n\mathbf{k}}(\mathbf{r}). \tag{6.24}$$

The back transformation is given by

$$\psi_{n\mathbf{k}}(\mathbf{r}) = \sum_{\mathbf{R}} e^{i\mathbf{k}\mathbf{R}} w_n(\mathbf{r}-\mathbf{R}), \tag{6.25}$$

where the sum is performed over all the unit cells in the crystal.

The Wannier functions obtained in Eq. 6.25 are not unique, because it is possible to mix the Bloch states of different band numbers by a unitary matrix $\mathbf{U}^{(\mathbf{k})}$. The resulting Wannier functions are also a complete representation of the electronic structure, although their localisation features are different:

$$w_n(\mathbf{r}-\mathbf{R}) = \frac{V}{8\pi^3}\int \mathrm{d}\mathbf{k}\, e^{-i\mathbf{k}\mathbf{R}}\left(\sum_m U^{(\mathbf{k})}_{mn}\psi_{m\mathbf{k}}(\mathbf{r})\right). \tag{6.26}$$

Since both transformations in 6.25 and 6.26 are unitary, and the original Bloch states are orthogonal, the resulting Wannier functions are also orthogonal.

6.1.4.2 Nonorthogonal Generalised Wannier Functions

The matrix $\mathbf{U}$ can be optimised in such a way that the resulting Wannier functions are maximally localised, as described in [2]. However, orthogonality and localisation are two competing properties, and more localised Wannier functions can be obtained if the orthogonality constraint is removed.

The linear combination of the Bloch orbitals of different bands can be performed by using a non-unitary matrix, resulting in nonorthogonal Wannier functions [3] $\phi_{\alpha\mathbf{R}}$:

$$\phi_{\alpha\mathbf{R}}(\mathbf{r}) = \frac{V}{8\pi^3}\int \mathrm{d}\mathbf{k}\, e^{-i\mathbf{k}\mathbf{R}}\left(\sum_m M^{(\mathbf{k})}_{m\alpha}\psi_{m\mathbf{k}}(\mathbf{r})\right). \tag{6.27}$$

In ONETEP, Wannier functions are constrained in a localisation sphere centred on atoms, i.e. $\phi_{\alpha\mathbf{R}} \equiv 0$ outside the localisation sphere, providing an atomic basis set. The radius of the localisation sphere is set by considering the electronic structure of the system or it can be increased until convergence of the physical properties is achieved. The nonorthogonal Wannier functions are optimised during the electronic structure calculation, hence they represent the "best possible" atomic basis functions of a particular system. In our studies of the atomic properties, we used these Wannier functions as the atomic basis set for calculating atomic properties with our definition for these properties given in Eq. 6.8.

6.1.5 Locality Investigations

In order to use the atomic energies obtained from quantum mechanical calculations as the target data of our regression scheme we have to ensure that the atomic energies are local. We tested the degree of this locality through the variation of the local energy caused by the perturbation of atoms outside a spatial cutoff. If the atomic energies are local, they can be regarded purely as functions of the local atomic environment and can be fitted by the Gaussian Process method.

The basic idea for testing the degree of the locality is that we generate a number of configurations, where the nearest neighbours of a certain atom were held fixed, while the positions of other atoms were allowed to vary. We calculated the local energy of the atom whose neighbourhood was fixed for each of these configurations and compared them. We then repeated this process for different neighbourhood configurations.

As a test system, we used clusters of 29–71 silicon atoms. The configurations were generated by molecular dynamics simulation at 3,000 K, where the forces were obtained from the Stillinger–Weber potential [4]. We performed the electronic structure calculations of the different clusters using ONETEP, and we also used ONETEP to determine the atomic energies, as described in Sect. 6.1.2. A typical cluster is shown in Fig. 6.1.

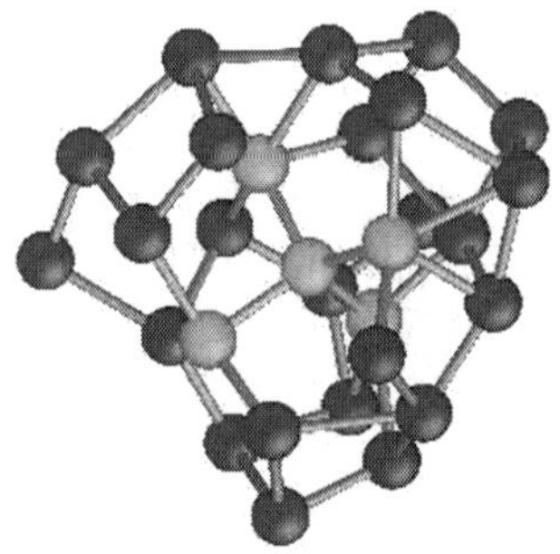

Fig. 6.1 An example configuration of the examined Si clusters. The atoms which were fixed during the molecular dynamics simulation are shown in a different shade

We examined the components of the atomic energies which depend principally on the electron density of the central atom. We calculated the average variation of the atomic kinetic, nonlocal and exchange-correlation energies and also, the total atomic energy corrected for the long-range interactions. The atomic energy was calculated as

$$E_i = Ekin_i + E_i^{\mathrm{nonloc}} + E_i^{\mathrm{xc}} + E_i^{ee} + E_i^{Ze} - \frac{1}{2}\sum_j^{\mathrm{atoms}} \hat{L}_i \hat{L}_j \frac{1}{r_{ij}}, \tag{6.28}$$

where the operator $\hat{L}$ can be written as

$$\hat{L}_i = q_i + \mathbf{p}_i \cdot \nabla_i + \mathbf{Q}_i : \nabla_i \nabla_i + \cdots. \tag{6.29}$$

The variations of the sum of kinetic, nonlocal and exchange-correlation terms are depicted in Fig. 6.2, while Fig. 6.3 shows the spread of atomic energies corrected for long-range interactions. Ideally, variations in the atomic energies should be within 0.1 eV for each neighbourhood as this is usually reckoned to be the standard DFT error. It is obvious that our results do not fit into this range. These results are not satisfactory and indicate that either the atomic energies depend on more neighbours, or that the atomic energies calculated by this particular method are not local. However, we found when using our final implementation of Gaussian Process (described in Sect. 4.6.1 in Chap. 4) an explicit definition of local energies is not necessary, as the Gaussian Process infers these from total energies and forces. We shall discuss the inferred atomic energies in Sect. 6.6.

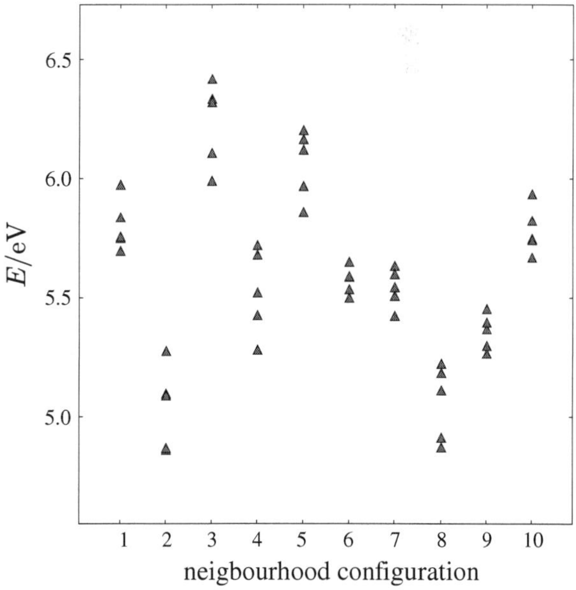

Fig. 6.2 Atomic energies $E = E_i^{\mathrm{kin}} + E_i^{\mathrm{nonloc}} + E_i^{\mathrm{xc}}$ of silicon clusters for different neighbourhoods

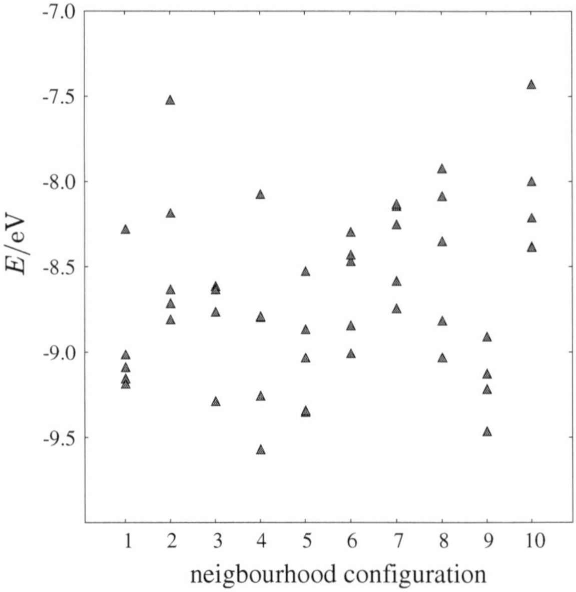

Fig. 6.3 Atomic energies of silicon clusters corrected for Coulomb contributions: $E = E_i - \frac{1}{2}\sum_j^{\text{atoms}} \hat{L}_i \hat{L}_j \frac{1}{r_{ij}}$ for different neighbourhood configurations

6.2 Gaussian Approximation Potentials

We have implemented the Gaussian Process to infer atomic energies from total energies and atomic forces. Gaussian Processes belong to the family of non-linear, non-parametric regression methods, i.e. not having fixed functional forms. The atomic environments are represented by the four dimensional bispectrum, which is invariant to permutation of neighbouring atoms and the global rotation of the environment. In order to demonstrate the power of this new tool, we built potentials for a few technologically important materials and we examined how closely the fitted potential energy surface is to the original, quantum mechanical one. At this stage of the work, most of the configurations we used for the training were close to the crystalline structure of the material, hence the use of the current potentials is limited to crystalline phases. However, to show the ability of our potential to describe mode widely varying configurations, in the case of carbon we built a potential that could describe the sp^2–sp^3 transition of the carbon atoms, the (111) surface of diamond and a simple point defect.

Our aim is twofold. On the one hand, we would like to generate potentials for general use, which can be extended, if needed. On the other hand, there are applications where "disposable" force fields are sufficient. For example, when simulating a crack or defects in a crystalline material, only a restricted part of the potential energy surface is accessible. In these cases, a purpose-built potential can be used, which can be generated more rapidly.

6.2.1 Gaussian Approximation Potentials for Simple Semiconductors: Diamond, Silicon and Germanium

Our first application of the Gaussian Approximation Potentials was a set of potentials for simple semiconductors. We calculated the total energies and forces of a number of configurations, which were generated by randomly displacing atoms in the perfect diamond structure. We included 8-atom and 64-atom supercells at different lattice constants and we perturbed the lattice vectors in some cases. The atoms were displaced at most by 0.3 Å.

The parameters of our representation are the spatial cutoff and the resolution of the bispectrum. We set the former to 3.7, 4.8 and 5.0 Å for carbon, silicon and germanium, respectively. The resolution of the bispectral representation can be changed by varying a single parameter, the maximum order $J_{\max}$ of the spherical harmonics coefficients we use when constructing the bispectrum. We used $J_{\max} = 5$ in all cases. During the sparsification, we chose 300 atomic neighbourhoods in all cases. Due to the method of generating these configurations all the neighbourhoods were similar, thus we decided to select the set of atomic environments for the sparsification randomly.

The electronic structure calculations were performed using CASTEP [5]. We used the local density approximation for carbon and the PBE generalised gradient approximation for silicon and germanium. The electronic Brillouin zone was sampled by using a Monkhorst–Pack k-point grid, with a k-point spacing of at most 1.7 $Å^{-1}$. The plane-wave cutoff was set to 350, 300 and 300 eV for C, Si and Ge, respectively, and the total energies were extrapolated for infinite plane-wave cutoff. Ultrasoft pseudopotentials were used with four valence electrons for all ions.

In Fig. 6.4 we show the performance of GAP, compared to the state-of-the-art interatomic potential, the Brenner potential [6]. The set of configurations used for testing was obtained from a long ab initio molecular dynamics run of a 64-atom supercell at 1,000 K. The absolute values of the components of the difference between the predicted and the DFT forces are shown as a function of the

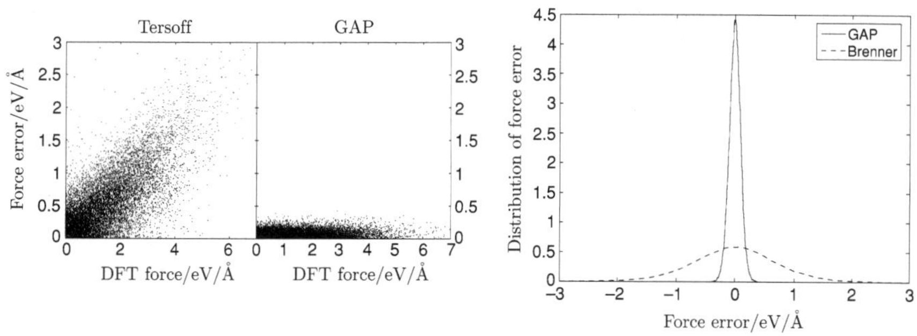

Fig. 6.4 Force errors compared to DFT forces for GAP and the Brenner potential in diamond. The *left* shows the force errors at different DFT forces. On the *right*, the distribution of the force errors is shown

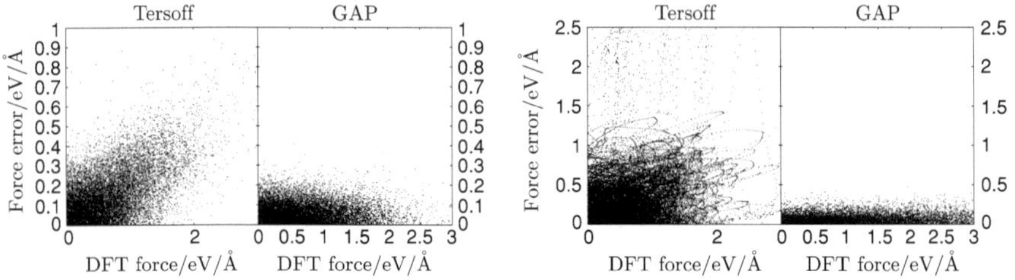

Fig. 6.5 Force errors compared to DFT forces for GAP and the Tersoff potential. The silicon potentials are shown in the *left* and the germanium potentials in the *right*

DFT force components and the distribution of these differences is also displayed. The force and energy evaluation with the Gaussian Approximation Potential for diamond, in the current implementation, is about 4,000 times faster than Density Functional Theory in the case of a 216-atom supercell.

We show in Fig. 6.5 the results for our potentials which were developed to model the two other group IV semiconductors, silicon and germanium, compared to the Tersoff potential.

The strict localisation of the atomic energies places a limit on the accuracy with which the PES can be approximated. If we consider an atom whose environment *inside* r_{cut} is fixed, but the position of other atoms are allowed to vary, the forces on this atom will still show a variation, depending on its environment *outside* the cutoff. An estimate of this theoretical limit can be obtained by calculating the force on an atom inside a fixed environment in various configurations. For carbon atoms in the diamond structure with $r_{cut} = 3.7$ Å this error estimate is 0.1 eV/Å.

6.2.2 Parameters of GAP

In diamond, we carried out the GAP training process using different parameters to determine the accuracy of the representation. We truncated the spherical harmonics expansion in Eq. 2.58 at J_{max}, which therefore represents the resolution of the bispectrum. Employing more spherical harmonics coefficients requires more computational resources, partially because of the increased number of operations needed for the calculation of the bispectrum and partially because there are more invariant elements, which affects the calculation of the covariances in Eq. 3.20. Figure 6.6 shows the force error of three different GAP models. The cutoffs of all three models were 3.7 Å, but the spherical harmonics expansion was truncated at the first, the third and the fifth channel, respectively. We chose $J_{max} = 5$ for our model, as in this case the standard deviation of the force errors reached the theoretical limit of 0.1 eV/Å associated with the spatial cutoff.

Figure 6.7 shows the force errors of three Gaussian Approximation Potential models for diamond with cutoffs of 2.0, 2.75 and 3.7 Å. The difference between the latter two models is negligible. However, the elastic moduli calculated from

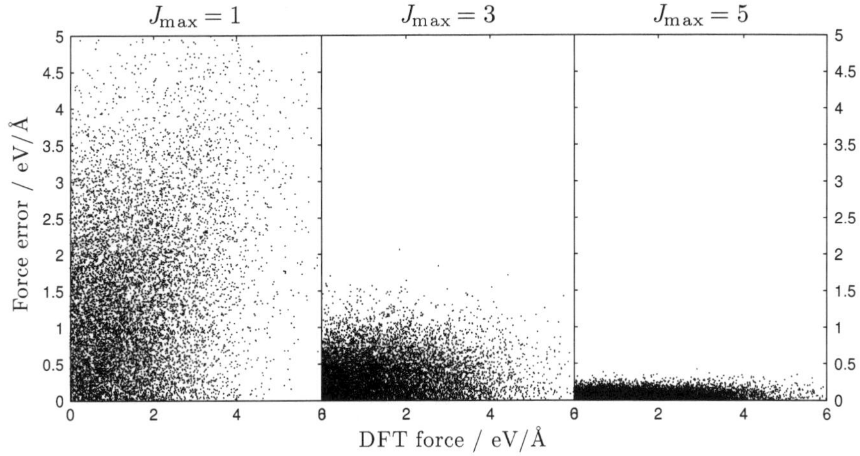

Fig. 6.6 Force correlation of GAP models for diamond with different resolution of representation. The number of invariants were 4, 23 and 69 for J_{max} = 1, 3 and 5, respectively

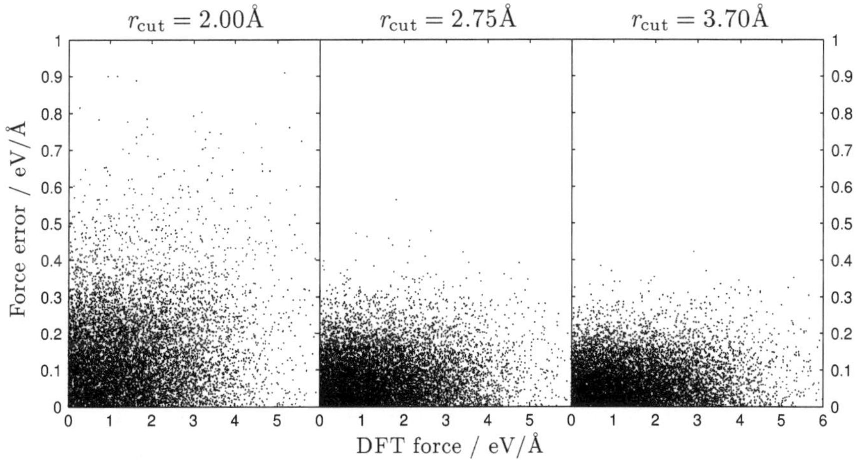

Fig. 6.7 Force correlation of GAP models for diamond with different spatial cutoffs

the model with r_{cut} = 2.75 Å did not match the elastic moduli of the ab initio model and so we chose r_{cut} = 3.7 Å for our final GAP potential.

6.2.3 Phonon Spectra

The force error correlation is already a good indicator of how well our potential fits the original potential energy surface. In addition, we determined the accuracy of a

Table 6.1 Parameters of the used GAP potentials

	C	Si	Ge
r_{cut}/Å	3.7	4.8	5.0
J_{max}	5	5	5

few other properties. The phonon dispersion curves represent the curvature of the potential energy surface around the lowest energy state. We calculated the phonon spectrum by the finite difference method using GAP. The force-constant matrix of the model was calculated by the numerical differentiation of the forces, and the phonon spectrum was obtained as the eigenvalues of the Fourier-transform of the force-constant matrix. The parameters of the GAP potentials are given in Table 6.1. We compared the phonon values at a few points in the Brillouin zone with the ab initio values and the analytic potentials. These results are shown in Figs. 6.8, 6.9 and 6.10 for diamond, silicon and germanium, respectively. The GAP models show excellent accuracy at zero temperature over most of the Brillouin zone, with a slight deviation for optical modes in the (111) direction. The agreement of the phonon spectrum of GAP with the phonon spectrum of Density Functional Theory suggests that any quantity that can be derived from the vibrational free-energy, such as the constant-volume heat capacity, at low temperatures will also show good agreement. We found excellent agreement between the phonon frequencies calculated by the GAP potential for diamond and the dispersion curves measured by inelastic neutron scattering [7, 8], shown in Fig. 6.11.

We also calculated the elastic constants of our models and these are compared to Density Functional Theory and existing interatomic potentials in Table 6.2. We note that to our current knowledge, no existing analytic potential could reproduce all of the elastic constants of these materials with an error of only a few percents.

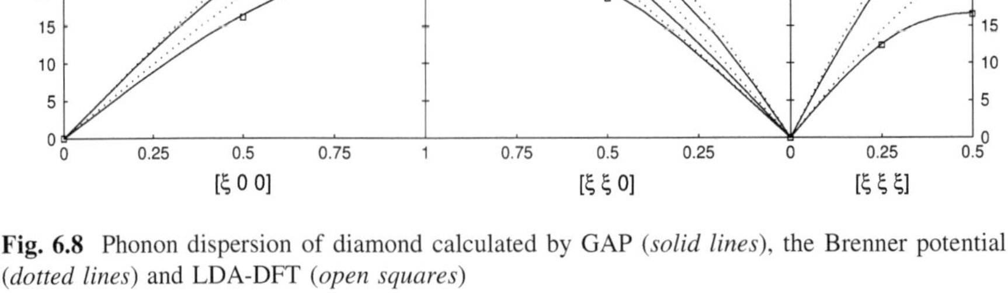

Fig. 6.8 Phonon dispersion of diamond calculated by GAP (*solid lines*), the Brenner potential (*dotted lines*) and LDA-DFT (*open squares*)

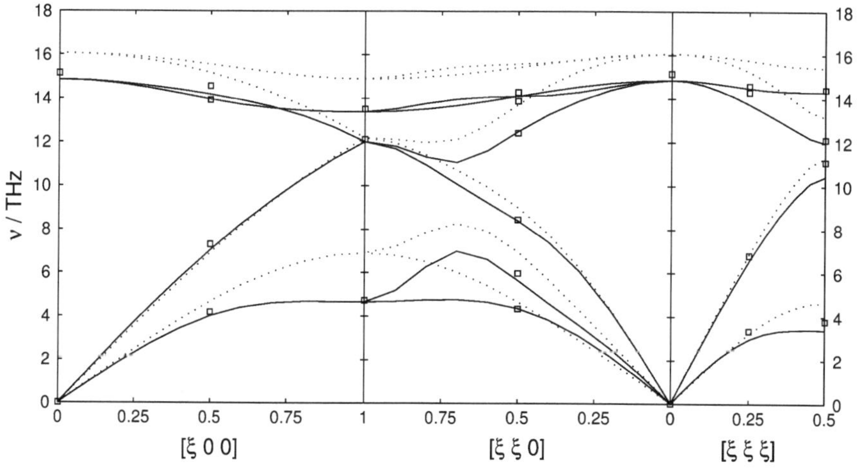

Fig. 6.9 Phonon dispersion of silicon calculated by GAP (*solid lines*), the Tersoff potential (*dotted lines*) and PBE-DFT (*open squares*)

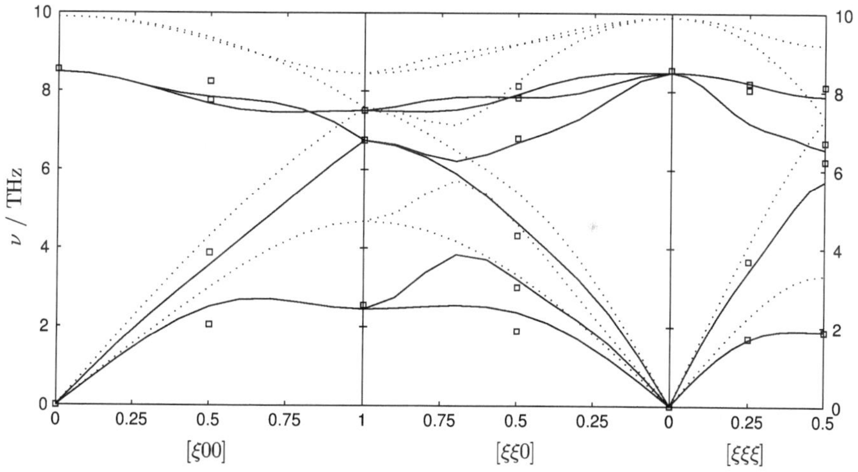

Fig. 6.10 Phonon dispersion of germanium calculated by GAP (*solid lines*), the Tersoff potential (*dotted lines*) and PBE-DFT (*open squares*)

6.2.4 Anharmonic Effects

In order to demonstrate the accuracy of the potential energy surface described by GAP outside the harmonic regime, we calculated the temperature dependence of the optical phonon mode of the Γ point in diamond. In fact, the low temperature variation of this quantity has been calculated using Density Functional Perturbation Theory by Lang et al. [9]. The ab initio calculations show excellent agreement

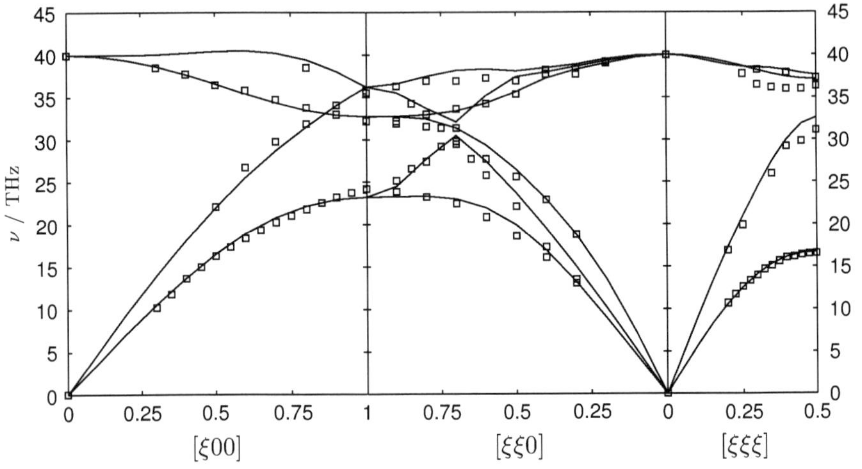

Fig. 6.11 Phonon dispersion of diamond calculated by GAP (*solid lines*) and experimental data points [7, 8] (*open squares*)

Table 6.2 Table of elastic constants, in units of GPa

	C		
	DFT	GAP	Brenner
C_{11}	1,118	1,081	1,061
C_{12}	151	157	133
C^0_{44}	610	608	736
C_{44}	603	601	717
	Si		
	DFT	GAP	Tersoff
C_{11}	154	152	143
C_{12}	56	59	75
C^0_{44}	100	101	119
C_{44}	75	69	69
	Ge		
	DFT	GAP	Tersoff
C_{11}	108	114	138
C_{12}	38	35	44
C^0_{44}	75	75	93
C_{44}	58	54	66

with experimental values determined by Liu et al. [10]. We calculated this optical phonon frequency using a molecular dynamics approach. We first performed a series of constant-pressure molecular dynamics simulations for a 250-atom supercell at different temperatures in order to determine the equilibrium lattice constant as a function of temperature. Then, for each temperature, we used the appropriate lattice constant to run a long microcanonical simulation, from which

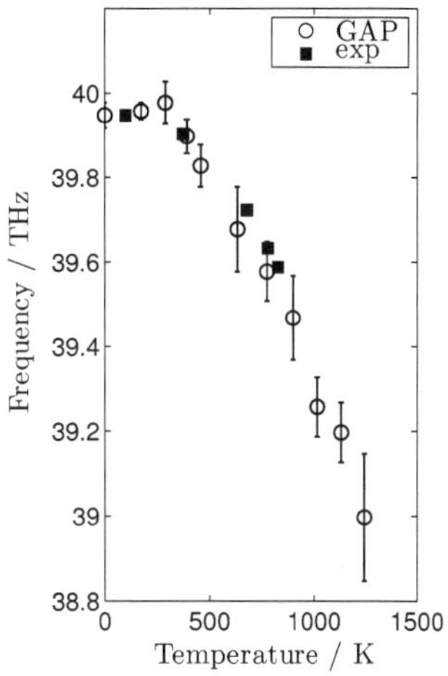

Fig. 6.12 Temperature dependence of the optical phonon at the Γ point in diamond

Table 6.3 Anharmonic shift of the Γ phonon frequency in diamond

	$\Delta\nu_{\text{anharmonic}}$/THz
LDA	0.95
GAP	0.93

we calculated the position–position correlation function. We selected the phonon modes by projecting the displacements according to the appropriate wavevector. From the Fourier-transform of the autocorrelation function, we obtained the phonon frequencies by fitting Lorentzians on the peaks. We present our results in Fig. 6.12, where our values for the phonon frequencies were shifted to match the experimental value at 0 K.

We note that even at 0 K there are anharmonic effects present due to the zero-point motion of the nuclei. We accounted for the quantum nature of the nuclei by rescaling the temperature of the molecular dynamics runs, by determining the temperature of the quantum system described by the same phonon density of states whose energy is equal to the mean kinetic energy of the classical molecular dynamics runs. The scaling function for the GAP model is shown in Fig. 6.13.

We are aware that at low temperatures this approximation is rather crude, and the correct way of taking the quantum effects into account would be solving the Schrödinger equation for the nuclear motion. However, we note that the anharmonic correction calculated by Lang et al. [9] by Density Functional Perturbation Theory and our value show good agreement (Table 6.3).

6.2.5 *Thermal Expansion of Diamond*

Another phenomenon that occurs as a result of the anharmonicity of the potential energy surface is thermal expansion. The temperature dependence of the thermal

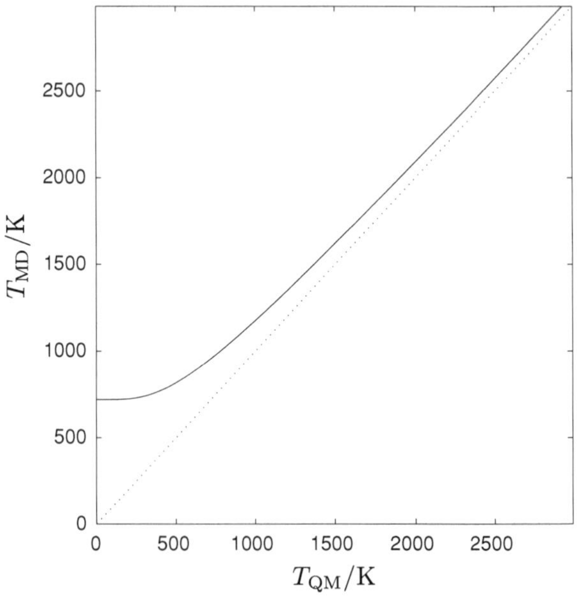

Fig. 6.13 Temperature of the quantum system described by GAP whose energy is equal to the average kinetic energy of the classical system, as a function of the temperature of the classical system. The *dotted line* is the identity function $f(x) \equiv x$, and is merely shown to provide a guide to the eye

expansion coefficient calculated from first principles using the quasi-harmonic approximation is remarkably close to the experimental value at low temperatures. However, at larger temperatures the quasi-harmonic approximation is less valid, because other anharmonic effects, which cannot be modelled assuming first-order dependence of the phonon frequencies on the lattice constant, are more significant. This effect can be calculated exactly by solving the nuclear Schrödinger equation for the nuclear motion, or by classical molecular dynamics simulation. Herrero and Ramírez used a path-integral Monte Carlo method to calculate the thermal expansion of diamond modelled by the Tersoff potential [11]. We determined the thermal expansion by calculating the equilibrium lattice constant by running a series of constant-pressure molecular dynamics simulations at different temperatures. We fitted the analytic function

$$a(T) = c_1 T + c_2 T^2 + c_{-1} T^{-1} + c_0 \tag{6.30}$$

to the lattice constants, and then calculated the thermal expansion using the definition

$$\alpha(T) = \frac{1}{a(T)} \left(\frac{\mathrm{d}a}{\mathrm{d}T} \right)_T . \tag{6.31}$$

The same analytic function was used by Skinner to obtain the thermal expansion coefficient from the experimental lattice constants [12]. Our results are shown in Fig. 6.14, together with the experimental values [12] and values calculated by LDA and GAP using the quasiharmonic approach. The results obtained by using the Brenner potential is shown in the right panel of Fig. 6.14. It can be seen that

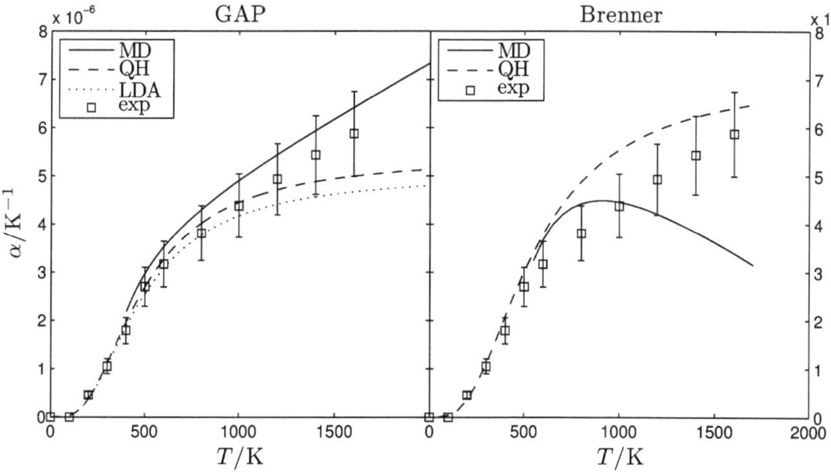

Fig. 6.14 Temperature dependence of the thermal expansion coefficient

the thermal expansion is extremely well predicted using GAP in molecular dynamics simulations.

The GAP results for the thermal expansion coefficients obtained from the quasiharmonic approximation show excellent agreement with the LDA values. This verifies that the potential energy surface represented by the GAP model is, in fact, close to the ab initio potential energy surface, even outside the harmonic regime. In the case of Density Functional Theory, the molecular dynamics simulation would be computationally expensive, because a large supercell has to be used to minimise finite-size effects. However, with GAP, these calculations can be easily performed and the thermal expansion coefficients obtained match the experimental values well, even at high temperatures.

6.3 Towards a General Carbon Potential

The ultimate aim of our research is to create potentials for general use. In the case of carbon, describing the diamond phase is certainly not sufficient. Although we still have to add many more training configurations to complete a general carbon potential, we demonstrate the capabilities of the GAP scheme by extending the scope of the diamond potential described in the previous section to include graphite, surfaces and vacancies.

We generated a set of randomised graphite configurations in a similar fashion to the diamond training configurations. We randomised the atomic positions of the carbon atoms in 54- and 48-atom supercells of rhombohedral and hexagonal graphite and we also considered a number of uniaxially compressed supercells. The training configurations also included diamond configurations with a vacancy

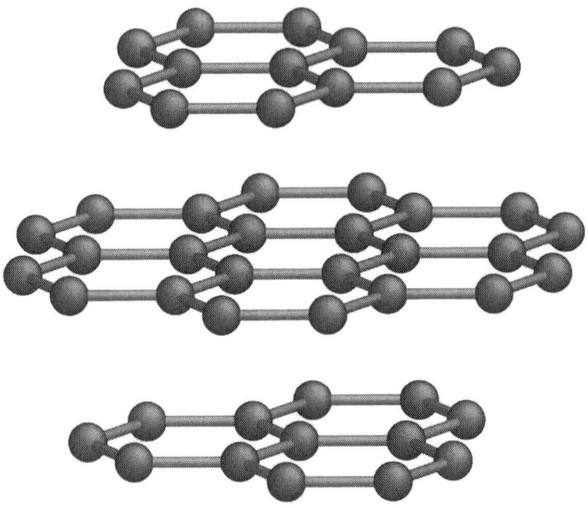

Fig. 6.15 Rhombohedral graphite

and (111) surfaces, in particular, configurations of the unreconstructed (111) surface and the 2 × 1 Pandey-reconstruction were included in the training set.

We tested how accurately the resulting GAP potential reproduces the rhombohedral graphite–diamond transition. Fahy et al. described a simple reaction coordinate that transforms the eight-atom unit cell of rhombohedral graphite (Fig. 6.15) to the cubic unit cell of diamond. In Fig. 6.16 we show the energies of the intermediate configurations between rhombohedral graphite and diamond calculated using GAP, DFT and the Brenner potential. The lattice vectors $\mathbf{a}_\alpha$ and the atomic coordinates $\mathbf{r}_i$ of these configurations were generated by

$$\mathbf{a}_\alpha = (1-x)\mathbf{a}_\alpha^{\text{graphite}} + x\mathbf{a}_\alpha^{\text{diamond}}, \quad \text{where } \alpha = 1,2,3 \tag{6.32}$$

$$\mathbf{r}_i = (1-x)\mathbf{r}_i^{\text{graphite}} + x\mathbf{r}_i^{\text{diamond}}, \quad \text{where } i = 1,\ldots,8. \tag{6.33}$$

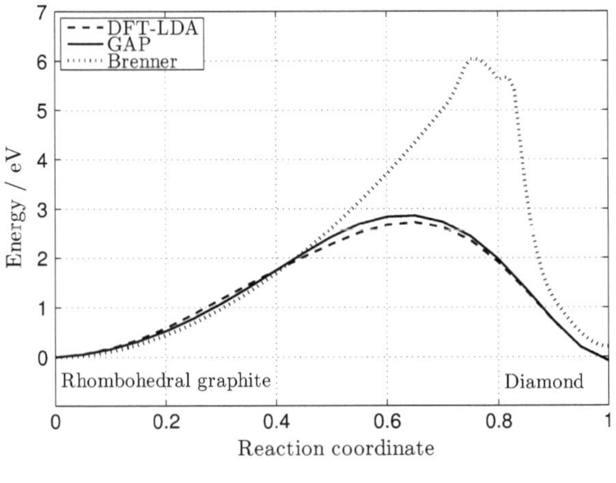

Fig. 6.16 The energetics of the linear transition path from rhombohedral graphite to diamond calculated by DFT, GAP and the Brenner potential

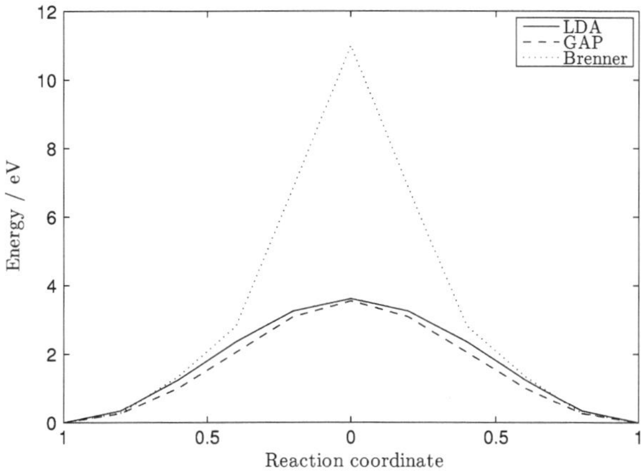

Fig. 6.17 Energy along a vacancy migration path in diamond by DFT, GAP and the Brenner potential

Table 6.4 Surface energies in the units of J/m^2 of the unreconstructed and 2×1 Pandey-reconstructed surface of the (111) diamond surface

	LDA-DFT	GAP	Brenner	Tersoff
unreconstructed	6.42	6.36	4.46	2.85
2×1	4.23	4.40	3.42	4.77

The reaction coordinate x corresponds to graphite at $x = 0$ and to diamond at $x = 1$. It can be seen that the Brenner potential cannot describe the change in the bonding of the carbon atoms, whereas the GAP potential reproduces the quantum mechanical barrier accurately.

We also calculated the energetics of the vacancy migration in a similar fashion, i.e. along a linear path between two configurations, where the vacancies are at two neighbouring lattice sites. Our results are shown in Fig. 6.17. The GAP model predicts the same the energies as the Density Functional Theory, whereas the Brenner potential overestimates the energy barrier of the migration.

Our results for the surface energies of the diamond (111) surface are presented in Table 6.4 again showing very good agreement between GAP predictions and LDA results.

6.4 Gaussian Approximation Potential for Iron

The Gaussian Approximation Potential scheme is not limited to simple semiconductors. We demonstrate this by applying the scheme to a metallic system, namely the body-centred cubic (bcc) phase of iron. We included configurations in the training set where the lattice vectors of the one-atom primitive cell were randomised and where the positions of the atoms in 8 and 16-atom supercells were also

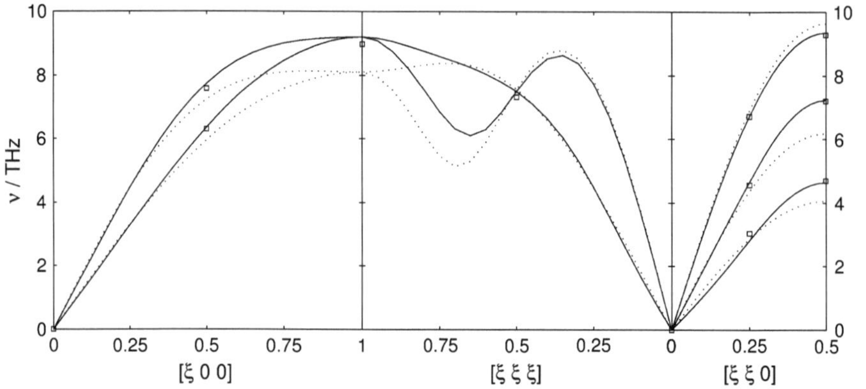

Fig. 6.18 Phonon spectrum of iron using the GAP potential (*solid lines*), the Finnis–Sinclair potential (*dotted lines*) and PBE-DFT (*open squares*)

randomised. These configurations were represented by 50 sparse points in the training set for the GAP potential. The spatial cutoff for the GAP potential was 4.0 Å and we used the spherical harmonics coefficients for the bispectrum up to $J_{\text{max}} = 6$.

We checked the accuracy of our potential by calculating the phonon spectrum along the high symmetry directions and comparing the phonon frequencies at a few k-points with Density Functional Theory. These spectra, together with those generated by the Finnis–Sinclair potential are shown in Fig. 6.18. In Fig. 6.19 we compared the phonon frequencies calculated by the GAP potential to the experimental values obtained by the neutron-inelastic-scattering technique [13]. The main features of the phonon dispersion relation, for example, the crossing of the two branches along the $[\xi, \xi, \xi]$ direction, are reproduced by the

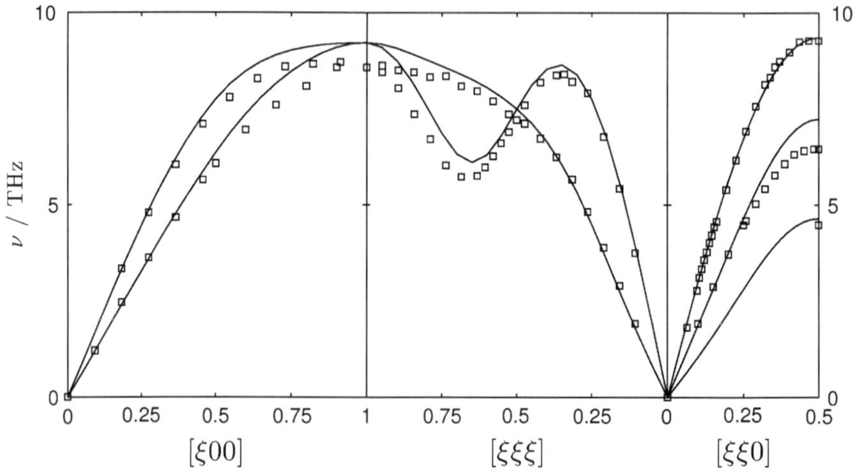

Fig. 6.19 Phonon dispersion of iron using the GAP potential (*solid lines*) and experimental values (*open squares*) [13]

Table 6.5 Elastic moduli of iron in units of GPa calculated using different models

	PBE-DFT	GAP	Finnis–Sinclair
C_{11}	236	222	245
C_{12}	160	156	138
C_{44}	117	111	122

GAP potential. The errors in the frequencies can be attributed to our Density Functional Theory calculations.

The elastic moduli calculated with our model, the Finnis–Sinclair potential [14] and Density Functional Theory are given in Table 6.5. The elastic properties and the phonon dispersion relations described by the GAP model show excellent agreement with the values calculated by Density Functional Theory.

6.5 Gaussian Approximation Potential for Gallium Nitride

So far our tests of the Gaussian Approximation Potentials were limited to single-species systems, but the framework can be extended to multispecies systems. Here we report our first attempt to model such a system, the cubic phase of gallium nitride. Gallium nitride (GaN) is a two-component semiconductor with a wurtzite or zinc-blende structure. There is a charge transfer between the two species.

As in our previous work, the configurations for fitting the GAP model were generated by randomising the lattice vectors of the primitive cell and randomly displacing atoms in larger supercells. Owing to the charge transfer, we need to include the long-range Coulomb-interaction in our model. We decided to use the charges obtained from the population analyses of the ground state electronic structure of a number of atomic configurations. Due to the fact that these configurations are similar, the fluctuation of the atomic charges was not significant, hence we chose to use a simple, fixed charge model with $-1e$ charge on the nitrogen atoms and $1e$ charge on the gallium atoms. We calculated the electrostatic forces and energies for each training configuration by the standard Ewald-technique [15] and subtracted these from the forces and energies obtained from the Density Functional Theory calculations. We regarded the remaining forces and energies as the short-range contribution of the atomic energies, and these were used for the regression to determine the GAP potential. The cutoff of the GAP potential was chosen to be 3.5 Å, $J_{\text{max}} = 5$ and we sparsified the training configurations using 300 sparse points.

We checked the correlation of the predicted forces of the resulting GAP potential with the ab initio forces, and the results are shown in Fig. 6.20. We used 64-atom configurations where the atoms were randomly displaced by similar amounts to the training configurations. The phonon spectrum calculated by GAP is shown in Fig. 6.21 and the elastic moduli are listed in Table 6.6.

Even this simple GAP model for gallium nitride shows remarkable accuracy in these tests, which we take as evidence that we can adapt GAP to multispecies

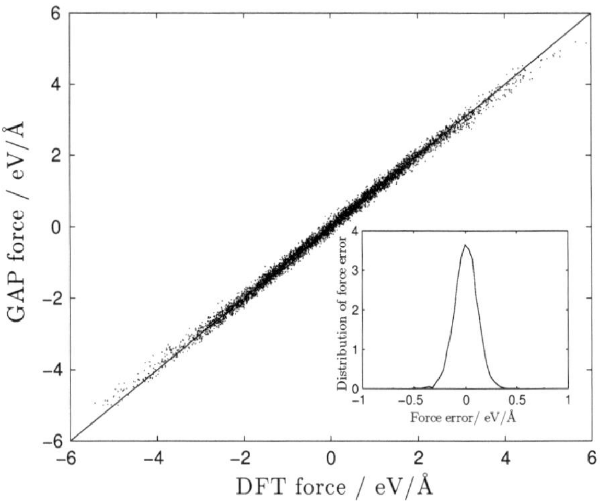

Fig. 6.20 Force components in GaN predicted by GAP versus DFT forces. The *diagonal line* is the $f(x) \equiv x$ function, which represents the perfect correlation. The *inset* depicts the distribution of the difference between the force components

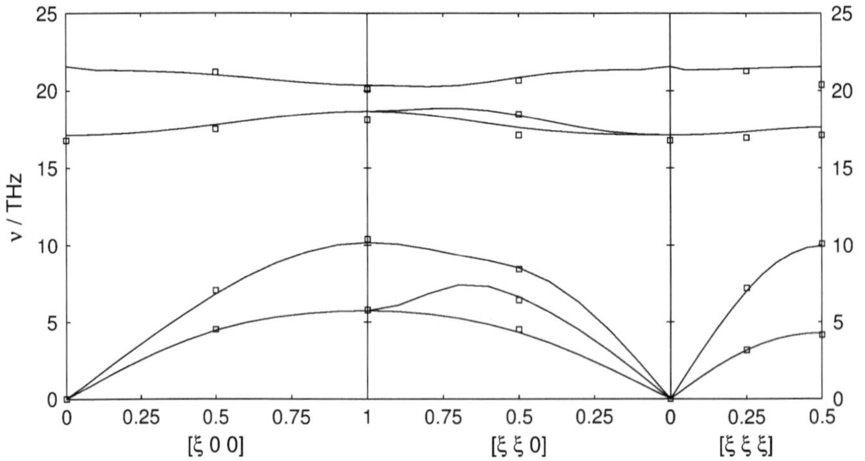

Fig. 6.21 Phonon spectrum of GaN, calculated by GAP (*solid lines*) and PBE-DFT (*open squares*)

Table 6.6 Elastic moduli of GaN in units of GPa calculated using PBE-DFT and GAP

	PBE-DFT	GAP
C_{11}	265	262
C_{12}	133	136
C_{44}	153	142

systems. However, in the case of very different neighbourhood configurations we will probably have to include variable charges, and we will possibly have to consider the contributions of multipole interactions in the long-range part of the potential. This is the subject of future research.

6.6 Atomic Energies from GAP

In Sect. 6.1.2 we investigated a possible definition of atomic energies based on localised atomic basis sets. According to our results in Sect. 6.1.5, however, those atomic energies could not be used in our potential generation scheme because they showed a large variation between numerically identical local environments. Instead, we employed some extensions of the Gaussian Process regression method—learning from derivatives, use of linear combination of function values and sparsification—, which make the explicit definition of atomic energies unnecessary. Nonetheless, we found it striking that an alternative possible definition of the quantum mechanical atomic energies, i.e. the ones inferred by the Gaussian Approximation Potentials appeared to be successful. In other words, using these atomic energies we can obtain the most commensurate forces and total energies for a given spatial cutoff, therefore these atomic energies are optimal in this sense.

We show two examples which demonstrate that the atomic energies predicted by GAP are consistent with physical considerations. In the first application, we calculated the atomic energies of the atoms in a 96-atom slab of diamond, which had two (111) surfaces. The training configurations were generated by scaling the lattice vectors and positions of the atoms of the minimised configuration by a constant factor and randomising the atomic positions, and each of these steps was started from a previous one. This means that in 20 steps, we created a series of samples between the minimised structure and a completely randomised, gas-like configuration. We calculated the total energy and the forces of the configurations by DFTB [16], and used these to train a GAP model. The cutoff of the model was 2.75 Å and the atomic environments were represented by 100 sparse teaching points. We used this model only to determine the atomic energies in the original slab. The atomic energies of the carbon atoms as a function of their distance from the surface are plotted in Fig. 6.22. It can been seen that the atomic energy is higher at the surface and then gradually reaches the bulk value towards the middle of the slab.

We also calculated the atomic energies defined by GAP in a gallium–nitride crystal where permutational defects were present. We created two configurations which contained such defects. The first one was generated by swapping the positions of a gallium and nitrogen atom in a 96-atom wurtize-type supercell, and then we swapped the positions of another pair to generate the second configuration. We calculated the total energies and forces of the two configurations by Density Functional Theory and used this data to train a very simple GAP potential. The cutoff of the model was 3.5 Å and we used six sparse point to represent the atomic environments. We used this model to calculate the atomic energies in the same two configurations. Certainly, the resulting potential is not a good representation of the quantum mechanical potential energy surface, but it still detects the defects and predicts higher atomic energies for the misplaced atoms.

Figure 6.23 shows the configurations with the defects and the perfect lattice. The colouring of atoms represent their atomic energies. It can be seen that the atomic energies of the atoms forming the defect and surrounding it are higher.

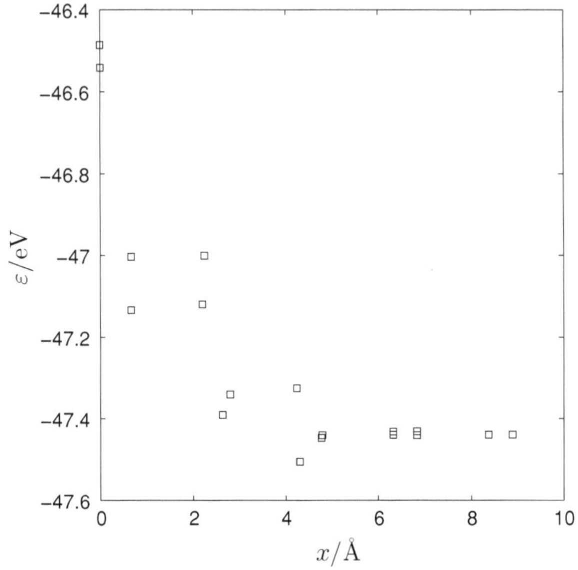

Fig. 6.22 Atomic energies of carbon atoms in a slab of diamond with two (111) surfaces as a function of the distance of the atom from the surface

In random structure search applications [17] GAP can be directly employed to detect permutational defects. If there are more than one species present in the structure, the structure search can result in many similar lattices, none of which are perfect, because of the large number of permutations of different species. GAP models, which are generated on the fly, can be used to suggest swaps of atoms between the local minima already found, which can then result in lower energy structure. Using GAP as an auxiliary tool in such structure searches can possibly achieve a significant speedup in searching for the global energy minimum.

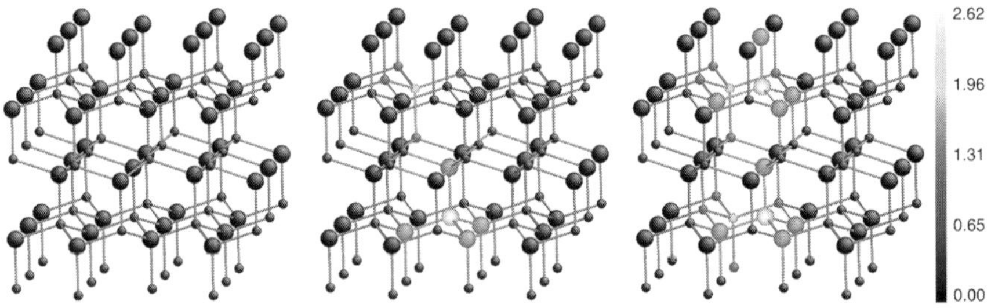

Fig. 6.23 The atomic energies in gallium–nitride crystals. We show the perfect wurtzite structure on the *left*, a crystal containing a single defect (a gallium and a nitrogen atom swapped) is on the *middle* and a crystal containing two defects (a further gallium–nitrogen pair swapped) is on the *right*. The smaller spheres represent the nitrogen atoms, the larger ones represent the gallium atoms. The *coloured bar* on the right shows the energy associated with the *colour shades*, in electrovolt

6.7 Performance of Gaussian Approximation Potentials

The total computational cost of Gaussian Approximation Potentials consists of two terms. The first term, which is a fixed cost, includes the computation of the ab initio forces and energies of the reference calculations and the generation of the potential. The time required to generate the potential scales linearly with the number of atomic environments in the reference configurations and the number of sparse configurations. In our applications, performing the DFT calculations typically took 100 CPU hours while the generation of a GAP potential was about a CPU hour.

Even for small systems, GAP potentials in our current implementation are order of magnitudes faster than Density Functional Theory, but significantly—about a hundred times—more expensive than analytical potentials. Calculation of the energies and forces requires about 0.01 s for every atom on a single CPU core. For comparison, a timestep of a 216-atom simulation cell takes about 190 s per atom on a single core by CASTEP, which corresponds to 20,000-fold speedup. The same calculation for iron would take a million times longer by CASTEP.

References

1. C.-K. Skylaris, P.D. Haynes, A.A. Mostofi, M.C. Payne, J. Chem. Phys. **122**, 84119 (2005)
2. N. Marzari, D. Vanderbilt, Phys. Rev. B **56**, 12847 (1997)
3. C.-K. Skylaris, A.A. Mostofi, P.D. Haynes, O. Diéguez, M.C. Payne, Phys. Rev. B **66**, 035119 (2002)
4. F.H. Stillinger, T.A. Weber, Phys. Rev. B **31**, 5262 (1985)
5. S.J. Clark et al., Zeit. Krist. **220**, 567 (2005)
6. D.W. Brenner et al., J. Phys.: Cond. Mat. **14**, 783 (2002)
7. J.L. Warren, J.L. Yarnell, G. Dolling, R.A. Cowley, Phys. Rev. **158**, 805 (1967)
8. J.L. Warren, R.G. Wenzel, J.L. Yarnell, *Inelastic Scattering of Neutrons*, (International Atomic Energy Agency, Vienna, 1965), p. 361
9. G. Lang et al., Phys. Rev. B **59**, 6182 (1999)
10. M.S. Liu, L.A. Bursill, S. Prawer, R. Beserman, Phys. Rev. B **61**, 3391 (2000)
11. C.P. Herrero, Phys. Rev. B **63**, 024103 (2000)
12. B.J. Skinner, Am. Mineral. **42**, 39 (1957)
13. V.J. Minkiewicz, G. Shirane, R. Nathans, Phys. Rev. **162**, 528 (1967)
14. M.W. Finnis, J.E. Sinclair, Phil. Mag. A **50**, 45 (1984)
15. P.G. Cummins, D.A. Dunmur, R.W. Munn, N.R.J, Acta Crystallogr. Sect. A **32**, 847 (1976)
16. E. Rauls, J. Elsner, R. Gutierrez, T. Frauenheim, Sol. Stat. Com. **111**, 459 (1999)
17. C.J. Pickard, R.J. Needs, Phys. Rev. Lett. **97**, 045504 (2006)

Chapter 7
Conclusion and Further Work

During my doctoral studies, I implemented a novel, general approach to building interatomic potentials, which we call Gaussian Approximation Potentials. Our potentials are designed to reproduce the quantum mechanical potential energy surface (PES) as closely as possible, while being significantly faster than quantum mechanical methods. To achieve this, we used the concept of Gaussian Process from Inference Theory and the bispectral representation of atomic environments, which we derived and adapted using the Group Theory of rotational groups.

I tested the GAP models on a range of simple materials, based on data obtained from Density Functional Theory. I built interatomic potentials for the diamond lattices of the group IV semiconductors and I performed rigorous tests to evaluate the accuracy of the potential energy surface. These tests showed that the GAP models reproduce the quantum mechanical results in the harmonic regime, i.e. phonon spectra, elastic properties very well. In the case of diamond, I calculated properties which are determined by the anharmonic nature of the PES, such as the temperature dependence of the optical phonon frequency at the Γ point and the temperature dependence of the thermal expansion coefficient. Our GAP potential reproduced the values given by Density Functional Theory and experiments.

These potentials constituted our initial tests of the scheme, and represented only a small part of the PES. In the case of carbon, I extended the GAP model to describe graphite, the diamond (111) surface and vacancies in the diamond lattice. I found that the new GAP potential described the rhombohedral graphite–diamond transition, the surface energies and the vacancy migration remarkably well.

To show that our scheme is not limited to describing monoatomic semiconductors, I generated a potential for bcc iron, a metal, and for gallium nitride, an ionic semiconductor. Our preliminary tests, which were the comparison of the phonon dispersion and the elastic moduli with Density Functional Theory values, demonstrate that GAP models can easily be built for different kinds of materials. I also suggest that the Gaussian Approximation Potentials can be generated on the fly and used as auxiliary tools for example, in structure search applications.

A. Bartók-Pártay, *The Gaussian Approximation Potential*, Springer Theses,
DOI: 10.1007/978-3-642-14067-9_7, © Springer-Verlag Berlin Heidelberg 2010

7.1 Further Work

In my thesis I presented preliminary tests and validation of our potential generation scheme. In the future, we intend to build models and perform large scale simulations on a wide range of materials. The first step will be to create a general carbon potential, which can describe amorphous and liquid carbon at a wide range of pressures and temperatures as well as defects and surfaces. We are also planning to create "disposable" potentials, which can be used, for instance, in the case of crack simulations. These do not have to be able to describe the high-temperature behaviour of the materials, as only a restricted part of the configurational space is accessible under the conditions of the simulation. The description of electrostatics will be soon implemented, with charges and polarisabilities which depend on the local environment and the electric field. This will allow us to simulate more complex systems, for example silica or water and our ultimate aim is to build interatomic potentials—force fields—for biological compounds. None of these potentials have to be based on Density Functional Theory, for instance it might be necessary to use more accurate solutions of the electronic Schrödinger equation. Finally, using GAP as a post-processing tool to determine atomic energies derived from on Quantum Mechanics is also a future direction of our research, for example, in structure searches.

Chapter 8
Appendices

8.1 A: Woodbury Matrix Identity

The likelihood function in Eq. 3.46 is used during the sparsification procedure in order to optimise the hyperparameters and the sparse points. At first sight, it seems that the inverse of an $N \times N$ matrix has to be calculated, the computational cost of which would scale as N^3. However, by using the *matrix inversion lemma*, also known as the Woodbury matrix identity, the computational cost scales only with NM^2 if $N \gg M$. If we want to find the inverse of a matrix, which can be written in the form $\mathbf{Z} + \mathbf{UWV}^T$, the Woodbury matrix identity states that

$$(\mathbf{Z} + \mathbf{UWV}^T)^{-1} = \mathbf{Z}^{-1} - \mathbf{Z}^{-1}\mathbf{U}(\mathbf{W}^{-1} + \mathbf{V}^T\mathbf{Z}^{-1}\mathbf{U})^{-1}\mathbf{V}^T\mathbf{Z}^{-1}. \tag{A.1}$$

In our case, $\mathbf{Z}$ is an $N \times N$ diagonal matrix, hence its inverse is trivial, and $\mathbf{W}^{-1}$ is $M \times M$. The order of the operations can arranged such that none of them requires more than NM^2 floating point operations:

$$\begin{aligned} &\mathbf{t}^T(\mathbf{C}_{NM}\mathbf{C}_M^{-1}\mathbf{C}_{MN} + \boldsymbol{\Upsilon})^{-1}\mathbf{t} \\ &= \mathbf{t}^T\boldsymbol{\Upsilon}^{-1}\mathbf{t} - (\mathbf{t}^T\boldsymbol{\Upsilon}^{-1})\mathbf{C}_{NM}(\mathbf{C}_M^{-1} + \mathbf{C}_{MN}\boldsymbol{\Upsilon}^{-1}\mathbf{C}_{NM})^{-1}\mathbf{C}_{MN}(\boldsymbol{\Upsilon}^{-1}\mathbf{t}), \end{aligned} \tag{A.2}$$

where $\boldsymbol{\Upsilon} = \boldsymbol{\Lambda} + \sigma^2\mathbf{I}$. In the evaluation of the second term in Eq. 3.46 we used the *matrix determinant lemma*, which is analogous to the inversion formula:

$$\det(\mathbf{Z} + \mathbf{UWV}^T) = \det(\mathbf{W}^{-1} + \mathbf{V}^T\mathbf{Z}^{-1}\mathbf{U})\det(\mathbf{W})\det(\mathbf{Z}). \tag{A.3}$$

In our implementation, the determinants are calculated together with the inverses, without any computational overhead.

We also note that at certain values of the hyperparameters the matrix $\mathbf{C}_M$ is ill conditioned. In the original Gaussian Process, the covariance matrix $\mathbf{Q}$ can also be ill conditioned, but by adding the diagonal matrix $\sigma_\nu^2\mathbf{I}$ this problem is eliminated, except for very small values of the σ_ν parameters. Snelson suggested[1] that a small

[1] E. Snelson, personal communication.

DOI: 10.1007/978-3-642-14067-9_8,

diagonal matrix $\xi^2\mathbf{I}$ should be added to $\mathbf{C}_M$ to improve the condition number of the matrix. This small "jitter" factor can be regarded as the internal error of the sparsification.

8.2 B: Spherical Harmonics

8.2.1 Four-dimensional Spherical Harmonics

The spherical harmonics in three dimensions are the angular part of the solution of the Laplace equation

$$\left(\frac{\partial^2}{\partial x^2}+\frac{\partial^2}{\partial y^2}+\frac{\partial^2}{\partial z^2}\right)f=0. \tag{B.1}$$

This concept can be generalised to higher dimensions. In our case, we need the solutions of the four dimensional Laplace equation

$$\left(\frac{\partial^2}{\partial x^2}+\frac{\partial^2}{\partial y^2}+\frac{\partial^2}{\partial z^2}+\frac{\partial^2}{\partial z_0^2}\right)f=0, \tag{B.2}$$

which can be written in the form of the three-dimensional rotation matrices, the Wigner D-functions.

The definition of the elements of the rotational matrices is

$$D^{(l)}_{mm'}(R)=\langle Y_{lm}|\hat{R}|Y_{lm'}\rangle, \tag{B.3}$$

where the rotation $\hat{R}$ is defined by three rotational angles. The rotational operator is usually described as three successive rotations

- rotation about the z axis by angle α,
- rotation about the new y' axis by angle β,
- rotation about the new z' axis by angle γ, where α, β and γ are called the Euler-angles. The Wigner D-functions are usually formulated as the function of these three angles and denoted as $D^J_{MM'}(\alpha,\beta,\gamma)$. However, in some cases the rotation can be described more conveniently in terms of ω, θ and ϕ, where the rotation is treated as a single rotation through angle ω about the axis $\mathbf{n}(\theta,\phi)$. The vector $\mathbf{n}$ is determined by the polar angles θ and ϕ.

The rotational matrices in the form $U^J_{MM'}(\omega,\theta,\phi)$, where the four dimensional polar angles are $2\theta_0\equiv\omega$, θ and ϕ are the four dimensional spherical harmonics.

The matrix elements can be constructed as

$$U^J_{MM'}(\theta_0,\theta,\phi)=\begin{cases}(-iv)^{2J}\left(\frac{u}{-iv}\right)^{M+M'}e^{-i(M-M')\phi}\\ \times\sum_s\frac{\sqrt{(J+M)!(J-M)!(J+M')!(J-M')!}}{s!(s+M+M')!(J-M-s)!(J-M'-s)!}(1-v^{-2})^s,\\ M+M'\geq 0\\ \\ (-iv)^{2J}\left(\frac{u^*}{-iv}\right)^{-M-M'}e^{-i(M-M')\phi}\\ \times\sum_s\frac{\sqrt{(J+M)!(J-M)!(J+M')!(J-M')!}}{s!(s-M-M')!(J+M-s)!(J+M'-s)!}(1-v^{-2})^s,\\ M+M'\leq 0\end{cases}, \tag{B.4}$$

where

$$v=\sin\theta_0\sin\theta \tag{B.5}$$

$$u=\cos\theta_0-i\sin\theta_0\cos\theta. \tag{B.6}$$

In our application, each time an entire set of $U^J_{MM'}$ has to be calculated, thus the use of recursion relation is computationally more efficient. The recursion relations are

$$\begin{aligned}U^J_{MM'}(\theta_0,\theta,\phi)&=\sqrt{\frac{J-M}{J-M'}}u^*U^{J-\frac{1}{2}}_{M+\frac{1}{2}M'+\frac{1}{2}}(\theta_0,\omega,\theta)\\&\quad-i\sqrt{\frac{J+M}{J-M'}}ve^{i\phi}U^{J-\frac{1}{2}}_{M-\frac{1}{2}M'+\frac{1}{2}}(\theta_0,\omega,\theta)\\&\quad\text{for } M'\neq J\end{aligned} \tag{B.7}$$

and

$$\begin{aligned}U^J_{MM'}(\theta_0,\theta,\phi)&=\sqrt{\frac{J+M}{J+M'}}uU^{J-\frac{1}{2}}_{M-\frac{1}{2}M'-\frac{1}{2}}(\theta_0,\omega,\theta)\\&\quad-i\sqrt{\frac{J-M}{J+M'}}ve^{i\phi}U^{J-\frac{1}{2}}_{M+\frac{1}{2}M'-\frac{1}{2}}(\theta_0,\omega,\theta)\\&\quad\text{for } M'\neq -J.\end{aligned} \tag{B.8}$$

The actual implementation does not involve the explicit calculation of the polar angles, we calculate the spherical harmonics in term of the Cartesian coordinates x, y, z and z_0. The first two four-dimensional spherical harmonics are

$$U^0_{00}=1 \tag{B.9}$$

and

$$U^{\frac{1}{2}}_{\pm\frac{1}{2}\pm\frac{1}{2}}=\frac{1}{\sqrt{2}}\frac{z_0\pm iz}{r} \tag{B.10}$$

$$U^{\frac{1}{2}}_{\pm\frac{1}{2}\mp\frac{1}{2}} = -\frac{i}{\sqrt{2}}\frac{x \mp iy}{r}, \tag{B.11}$$

which are indeed analogous to their three-dimensional counterparts.

8.2.2 Clebsch–Gordan coefficients

We used the following formula to compute the Clebsch–Gordan coefficients:

$$\begin{aligned} C^{c\gamma}_{a\alpha b\beta} &= \delta_{\gamma,\alpha+\beta}\Delta(abc) \\ &\times \sqrt{(a+\alpha)!(a-\alpha)!(b+\beta)!(b-\beta)!(c+\gamma)!(c-\gamma)!(2c+1)} \\ &\times \sum_z \frac{(-1)^z}{z!(a+b-c-z)!(a-\alpha-z)!(b+\beta-z)!(c-b+\alpha+z)!(c-a-\beta+z)!}, \end{aligned} \tag{B.12}$$

where Δ-symbol is

$$\Delta(abc) = \sqrt{\frac{(a+b-c)!(a-b+c)!(-a+b+c)!}{(a+b+c+1)!}}. \tag{B.13}$$